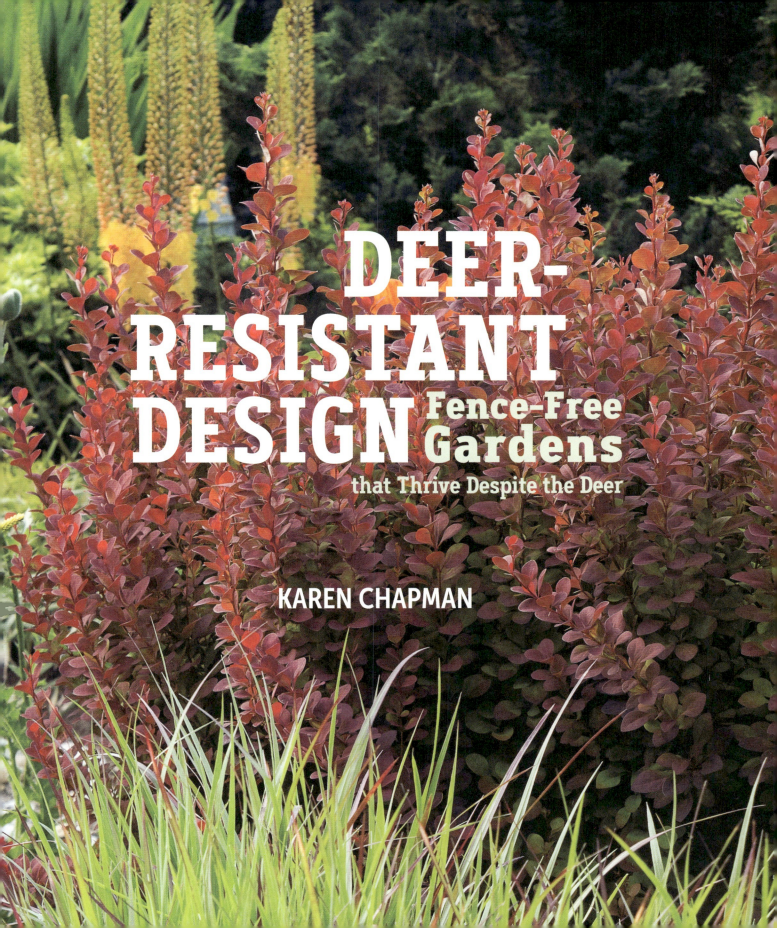

Frontispiece: The author's garden.

Copyright © 2019 by Karen Chapman. All rights reserved.
Illustrations copyright © 2019 by Kaitlin Pond.
Photo credits appear on page 232.

Published in 2019 by Timber Press, Inc.,
a subsidiary of Workman Publishing Co., Inc.,
a subsidiary of Hachette Book Group, Inc.
1290 Avenue of the Americas
New York, NY 10104

timberpress.com

Printed in China on responsibly sourced paper
Fourth printing 2023
Text and cover design by Mary Winkelman Velgos

The publisher is not responsible for websites (or their content) that are not owned by the publisher.

The Hachette Speakers Bureau provides a wide range of authors for speaking events. To find out more, go to hachettespeakersbureau.com or email HachetteSpeakers@hbgusa.com.

ISBN 978-1-60469-849-7

Catalog records for this book are available from the Library of Congress and the British Library.

To Anna Marjorie, the next generation.
Nana xx

Contents

Preface 8
Introduction: Strategies and Realities 10

Designer's Dream Garden—despite wildlife, water, and weeds 18
Country Garden—where daunting challenges become exciting opportunities 38
Desert Oasis—where neither heat, nor drought, nor deer diminish the mirage 52
Blue Jeans Garden—where kids and dogs are always welcome 64
Storyteller's Garden—unveiling magic, one chapter at a time 78
Garden of Survivors—where serendipity and inspired design go hand in hand 96
Garden of Connections—between people, flora, and fauna 108
Confetti Garden—where pops of color punctuate the landscape 124
Lake House Garden—where every day is a vacation 138
Collector's Garden—where hostaholics and deer draw a tentative truce 152
Hilltop Hacienda—where native and adapted plants bring Mexico to Texas 164
Curated Garden—where everyday natives and choice ornamentals find common ground 176
Suburban Retreat—bringing Colorado to New Jersey 192
Deer-Resistant Container Gardens 206

Resources 228
Metric Conversions 230
Further Reading 230
Acknowledgments 231
Photo Credits 232
Index 233

PREFACE

Gardeners who share their space with deer are constantly frustrated as they try to emulate the glossy magazine images they see of beautiful landscapes—and quickly realize that the majority of those tantalizing plant suggestions and combinations would read like a tasty menu to local herds. It's a problem I understand all too well and one that prompted me to write this book.

As I surveyed the stubby remains of my favorite coral bells (*Heuchera* 'Peach Flambé'), I realized that despite a mere fifteen miles of separation, gardening in rural Duvall, Washington, was going to be very different from my previous gardening experiences in suburban Kirkland, where deer were not a problem. Even with shelves straining under the weight of excellent reference books and stacks of gardening magazines to scan for ideas, I encountered only an occasional article on designing a garden that deer would not devour, or an uninspiring list of suggested plants which, although deer-resistant, were seldom suitable for my sticky clay soil and wet winter climate. I needed ideas for expansive borders as well as more intimate vignettes, specimen trees and focal points, bold sweeps of color and intriguing textures, and as a designer, I wasn't willing to settle for anything less than fabulous.

If this sounds familiar, take heart. This book has been written to encourage and inspire homeowners just like you with stories and photographs of mature landscapes that have withstood the test of time *and* the taste-testing of deer. It showcases gardens across the United States that successfully co-exist with wildlife, gardens whose owners have not resorted to fences or pouring concrete in despair. Each chapter features the story of one garden and the tenacious gardeners who tend it, describing their personal design criteria, the challenges they face, and the strategies they employ as they refuse to be thwarted. While a Portland couple enjoys a "live and let live" approach, delighting in observing the large herds of deer and elk that share their hilltop farm, they minimize noticeable browsing by planting in large drifts and selecting primarily deer-resistant plants. An avid hosta collector in New York proves that a regular spraying regime using a homemade concoction really does keep the deer out of his lush garden. (I had to see this garden myself to believe it!) A busy young couple in Michigan designed a Blue Jeans Garden, a wonderfully casual space that rarely suffers damage from passing deer since they observed the routes the herd used when traversing their land and chose the toughest, most deer-resistant shrubs for those areas. The common thread through all thirteen gardens is the determination of the homeowners, who prove that working with a restricted plant palette doesn't mean compromising on beauty or vision.

Whether your preferred design style is informal or elegant, whether you need a waterwise design or one that will withstand brutal winters, even if you are determined to grow hundreds of hostas (aka "caviar" to deer), this book has tales to tell and knowledge to impart. All the homeowners featured in these pages also share their "Top 10" deer-resistant plants, a welcome addition to a deer-challenged gardener's shopping list and guaranteed to put the fun back into nursery visits. Finally, a chapter on deer-resistant container gardens provides suggestions for making your patio pots every bit as colorful, captivating, and imaginative as those your city-living friends enjoy, with styles that run the gamut from contemporary to country, and exuberant to demure.

Landscape architects, designers, master gardeners, garden consultants, and nursery professionals will also benefit from the resources this book provides. In preparation for an appearance on PBS's *Central Texas Gardener*, I decided to visit a well-stocked nursery in Austin to familiarize myself with the local plant selection. I happened to overhear a customer asking a member of staff what he could plant that would be deer-resistant. The staff member smiled: "Well, they don't usually eat sage." The gentleman explained that he already had that—what else could he plant? "Oh, deer will eat anything if they're hungry enough," she said—and walked away. I was shocked. Her response was certainly true, but it was hardly helpful. There were so many things she could have offered, and questions she should have asked, starting with how much sun or shade the area he wished to plant received. So I introduced myself and offered to help. We then spent a very enjoyable and productive time strolling around the nursery, selecting a collection of colorful shrubs for his partial-shade garden—ones that looked fabulous together—by considering foliage color, texture, and size as well as deer resistance. The customer left smiling, delighted to realize that he really could have a garden to be proud of, despite the deer. I can't recall the gentleman's name now; but sir, should you happen to read this, know that I have written this book for you and customers like you, as well as for that nursery professional, who clearly was in need of ideas herself.

INTRODUCTION
Strategies and Realities

As the oft-repeated adage states, deer really will eat anything if they are hungry enough, but there is more to it than that, and understanding deer behavior can give us insights into their habits, thereby helping us stay one step ahead.

With no dogs at home, the deer became increasingly bold.

Deer are social learners, fawns accompanying their mothers to feeding grounds, often bypassing native forests to visit the same gardens from one year to the next. The availability of wild foods appears to have a major role in determining the level of browsing observed in the home landscape. Foraged acorns and mushrooms, two favorite, nutritious foods, are limited to certain seasons; and in the forest understory, plantings may be browsed so heavily that they are no longer within reach, even if a deer stands on its hind legs. Hostas and daylilies, on the other hand, are available for up to nine months of the year and can always be relied upon to be within snacking distance for even the youngest fawns.

Three-month-old Molly is now in training to chase the deer.

Many gardeners begin using deer repellent sprays in late winter, having observed that if deer can be taught very early in the year that a plant is unpalatable they will move away. Timing and constancy are crucial to success, however. If the plants taste delicious initially and *then* the plant is sprayed, the deer are more likely to return to try again.

In the United States, the major predator of deer is the coyote, and since dogs smell like coyotes the presence of domesticated dogs can be a deterrent. Reward your canine companion for alerting you to the presence of deer or, better still, for chasing them off your property; it's an easy way to give your pup some exercise and will save you from running out in slippers and bathrobe, waving your arms like a windmill while the deer continue to nonchalantly chew your tulips. Encouraging your dog to walk the perimeter of the garden and perform his bathroom duties in those areas can also significantly reduce the number of deer that enter your garden, since they perceive the smell of dog urine and feces as the presence of a predator.

I have certainly found a strong correlation between the boldness of deer and the presence of dogs. When we had dogs living with us and sharing our garden, the deer caused significantly less damage, and they kept farther away from the house. When our last adult dog passed away almost two years ago, the deer began virtually knocking on the door for their lunch. We now have a golden retriever puppy in training for future deer duty.

Deer do more than just eat foliage and flowers, however. The damage caused by bucks rubbing their velvety antlers against tree bark or spiky plants such as yucca can be significant and is all the more frustrating when these plants are otherwise considered to be deer-resistant. According to wildlife specialist Ken Gee, both rubbing and scraping behaviors are associated with the fall mating season, itself referred to as the rut. The effects of these two typically nocturnal behaviors serve as scent and possibly visual signposts between deer. Rubbing, when the buck rubs his antlers and forehead on a shrub or small tree, often starts shortly after the velvet has dried; this deposits scent from the forehead gland on the bark and creates a worn area known as a rub, which may then be recognized and used by several bucks or does. Several weeks after the first rubs appear, the second pattern of behavior, scraping, becomes evident; this is exemplified by a buck

Damage caused by deer rubbing their antlers on a young Leyland cypress (*Cupressus* ×*leylandii*).

An unobtrusive way to keep deer out of the garden: fishing wire works for New York gardener Mike Shadrack.

pawing a patch of ground and urinating in that area. The scrape is often adjacent to a low branch that has been broken by the buck biting or pulling on it. Scent from the forehead, preorbital gland (in front of the eye), or mouth is usually deposited on the broken branch, creating additional olfactory markers. Many homeowners featured in this book share their methods for protecting vulnerable garden plants during this challenging rutting season.

When Is a Fence Not a Fence?

The only foolproof way to keep deer from eating plants in your garden is to deny them entry in the first place, and while some gardeners might install tall fences to exclude deer from their properties, for many homeowners this isn't an option. Prohibitive cost, fence height restrictions, and hazy rural boundaries all make fencing a significant challenge, not to mention that few homeowners want to feel as though they live in a fortress. Our own five-acre garden has a wooded perimeter for the most part, and we have often joked that knowing our luck we would fence the deer *in*, as they have so many places to hide!

I witnessed some ingenious fencing solutions as I scouted gardens for this book, however. Mike Shadrack is president of the Western New York Hosta Society and owner of Smug Creek Gardens, a wooded property of some thirteen acres. To keep deer from eating his prized hostas, Mike uses multiple parallel lines of fishing wire strung between tree trunks to create an almost invisible fence. Occasionally deer penetrate the barrier and Mike has to repair the breach, but in general the deer are kept out. This

LEFT A lattice of rebar keeps deer out of Barbara and Howard Katz's garden in Bethesda, Maryland.

ABOVE Cedar posts create an informal deer-proof barrier in Tait Moring's garden in Austin, Texas.

OPPOSITE ABOVE A cattle grid makes it difficult for cloven-hooved cows—or deer—to enter this Georgia property.

OPPOSITE RIGHT In Austin's climate, vines quickly smother the posts, disguising their rugged appearance and creating a lush backdrop for bright flowers and containers.

method of exclusion might work for others, especially if the deer pressure is not too great.

Another remarkably inconspicuous deer fence was installed in the Maryland garden of designer Barbara Katz when her husband, Howard, discovered with horror that deer had entered from the neighboring property. Frequently on tours and photographed for major magazines, this garden had not been designed with deer in mind, so emergency measures needed to happen and fast! Howard came to the rescue by designing a sturdy lattice of rebar, which was attached to the existing concrete retaining wall. Now rusted, the entire structure is invisible from the home, melding discreetly into the shrubbery.

Landscape architect Tait Moring uses his garden as a laboratory in which to experiment with plants as well as an inspirational venue for clients. Besides choosing deer-resistant plants, Tait uses a deep stone wall at the garden's gated entrance to prevent deer from gaining access; he also uses cedar posts of various heights to exclude deer from the garden. "It works well until someone leaves the gate open," he laughs.

A cedar arch in Tait Moring's garden adds height to the entry gate, which connects to the cedar fence, densely clothed with vines.

Ingenuity is not limited to fence design, however. Where gates could not be installed across a driveway, an enterprising homeowner in Atlanta installed a cattle grid to keep deer from entering the garden.

Deer-Resistant Plants

Trying to outwit the deer is a lifelong challenge for those of us who live in deer country. Just when we think we understand their preferences, they develop a taste for something new. I personally suspect they can read price tags, as they always seem to eat the most expensive specimens.

When exclusion isn't an option, gardeners need to choose their plants wisely, aiming to incorporate only those that are reliably deer-resistant. Most experts will tell you that deer are less tempted by plants that contain poisonous compounds, or those with sharp thorns or aromatic or fuzzy foliage. If only the deer would read that memo. While I do agree these are good general guidelines, in my own garden I have witnessed deer chomping on spiky barberries as well as a highly toxic castor oil plant (*Ricinus communis* 'Carmencita'), and thorns certainly do not deter them from eating the roses. It is, however, a place to start.

No plant is entirely deer-proof, and plants vary considerably in their degree of deer resistance, from those that are rarely nibbled to others that suffer frequent damage. Adding to the complexity, the level of resistance may vary with season, availability of other food sources, the age of the deer, weather conditions, and the herd itself. While researching this book, I came across numerous instances where a plant was reliably deer-resistant in one garden but suffered damage in another just a few miles away. From state to state the differences were even more marked—cultivars of heavenly bamboo (*Nandina domestica*) being eaten to the ground in Washington yet considered a deer-resistant shrub in parts of Texas, for example.

In an effort to make sense of all this and provide meaningful, helpful information to homeowners, I have adopted the rating system used by the Rutgers New Jersey Agricultural Experiment Station (NJAES), which identifies four levels of deer resistance:

A = Rarely Damaged

B = Seldom Severely Damaged

C = Occasionally Severely Damaged

D = Frequently Severely Damaged

Their website (njaes.rutgers.edu/deer-resistant-plants) includes a searchable list of landscape plants, each rated for deer resistance. This list was compiled with input from nursery and landscape professionals, NJAES and Cooperative Extension personnel, and Master Gardeners in northern New Jersey. They recommend that plants in categories A and B would be best for landscapes prone to deer damage. Plants in categories C and D should be planted only with additional protection, such as the use of repellents.

In this book, when citing deer-resistance ratings for plants that were not included in Rutgers' list, I

COMMON PLANT TERMS

ANNUAL A plant that completes its growing cycle in one season and then dies. Typically, this includes any plant that would survive winters in zone 10 or warmer, even though it may be suitable as a houseplant or thrive in more tropical conditions.

PERENNIAL A plant that may be deciduous or herbaceous but comes back each year.

HERBACEOUS A plant that dies to the ground in winter but re-emerges in spring.

DECIDUOUS A plant that loses its leaves in fall but retains its twiggy structure through winter.

SEMI-EVERGREEN A plant that may lose a portion of its leaves in prolonged cold winters but will grow fresh foliage in spring.

EVERGREEN A plant that keeps its leaves year-round.

relied upon the knowledge and experience of local professionals and respected horticultural organizations. In cases where I witnessed considerable variation between the rating quoted and my own observations, I have noted the range.

Finally, it is important to understand the cultural requirements of the deer-resistant plants you have selected, to ensure they will thrive. The following terms are used throughout the book and indicate how much sun or shade a plant needs.

Full sun. At least six hours of direct sun each day.

Partial sun. Between four and six hours of direct sun, with protection during the hottest part of the day.

Partial shade. Morning sun only.

Full shade. Less than two hours of direct sun with some filtered sun during part of the day.

The USDA hardiness rating system provides gardeners a way to determine which plants will survive in their geographical location based on the average annual frost-free days and minimum winter temperatures. The lower the zone number, the colder the winter temperatures. This system has been used throughout the book to indicate the lower and upper zones within which a plant is hardy. To see temperature equivalents and to learn which zone you garden in, visit planthardiness.ars.usda.gov/PHZMWeb/. For Canada, go to planthardiness.gc.ca. For the UK, search for "hardiness" at rhs.org.uk.

Strategies and Realities 17

A Designer's Dream Garden
despite wildlife, water, and weeds

Rising from a congested tangle of brambles, the dead maple tree struck an imposing silhouette against the sky. Some homeowners might find such a vision intimidating. To me it was the most promising feature in our new garden. "We just need to clear the weeds before we can start planting," I remarked casually to my husband. Little did I know . . .

After months of searching, Andy and I believed we had finally found the perfect property in rural Duvall, thirty miles east of Seattle in the foothills of the Cascade Mountains. The modest single-story house and barn sat well back from the gravel road, surrounded by level pasture, a grassy meadow, and a forest of towering conifers interspersed with native shrubs whose names I was yet to learn. Swallows dipped and dived as we sauntered around the tranquil five acres on a perfect summer's day,

QUICK FACTS
LOCATION: Duvall, Washington (zone 6)
SOIL TYPE: clay
PROPERTY SIZE: 5 acres
PROBLEM CRITTERS: deer, rabbits, voles, lazy barn cats
OTHER CHALLENGES: seasonal flooding; summer drought; no irrigation

DESIGN CREDITS:
home of Andy and Karen Chapman
designed by Karen Chapman, Le Jardinet

dreaming of how we would renovate the outdated home and transform the landscape into our private oasis. With Andy's woodworking, engineering, and DIY skills and my horticultural and landscape design background, anything seemed possible. Teamwork, a sense of humor, and tenacity helped Andy and me hold true to the vision we formed that day.

We moved in on 30 October 2009, camping in our new home with only the essentials, storing most of our furniture and possessions in the barn while the builders set about the crucial rewiring, replumbing, and restructuring that were needed to make the dated home safe and secure. Those first six weeks were a series of daunting discoveries: a chorus of singing frogs in the crawl space (the bath tub drained directly onto the dirt floor beneath the house), dead mice in the walls (together with several pounds of birdseed), and electrical wiring that consisted of scraps of bare copper wire twisted loosely together to span the distance between receptacles and junction boxes. It's a miracle the house was still standing. Nevertheless, our dry British humor and sense of adventure remained mostly intact—until it started to rain.

Fall was exceptionally wet that year, even for Seattle, and we watched with mounting dismay as our home became surrounded by a five-acre moat. That was my first clue that we had clay soil. My romantic dream of strolling through our idyllic country garden was swiftly replaced by the reality of squelching through muddy, ankle-deep pools that I knew would be teeming with man-eating mosquitoes come spring. How could I grow anything at all with the land virtually submerged for several months of the year? Bitter disappointment threatened to overwhelm me as I surveyed the disheartening scene, and I began to wonder if we had made a terrible mistake.

Thank goodness for friends. I was urged to seek the help of John Silvernale, an experienced landscape architect with Berg's Landscaping, to see if the drainage problems could be corrected. Undaunted, he designed a network of 3-foot-deep French drains penetrating the sticky clay soil and inserted perforated pipes to carry excess ground water away, while enabling most of the rain to percolate into the surrounding soil. Finally, I could start to create my dream garden. I began planting in earnest in the spring of 2012, using Rutgers' list of deer-resistant plants as my guide. Even though it was compiled with input from nursery and landscape professionals in northern New Jersey—a long way from Seattle—it seemed like a good place to start.

The previous homeowners had warned us that they saw two or three deer in the garden most days, but we didn't see any until that fall, when seemingly overnight young conifers were ravaged, sedums were ruthlessly munched, and saplings suffered major bark wounds. I could have cried. I began to realize that plants being eaten was only part of the problem. I was witnessing significant damage by male deer rubbing their antlers and forehead against

Inspiration for the location of a showcase border came from inauspicious beginnings—a dead bigleaf maple (*Acer macrophyllum*).

Casual dining is easy when the patio is close to the kitchen as well as the vegetable garden.

tree bark during rutting season. Where some gardeners might install tall fencing around their landscape to keep deer out, this wasn't an option for us, due both to budget constraints and to the somewhat hazy property boundaries long since overgrown with woodland trees and brush. Knowing my luck, we would have fenced the deer in, along with the black bears, cougars, bobcats, and coyotes, all of which have since been seen. Meanwhile the rabbit population was exploding and seemed hellbent on eating the choice ornamental grasses and coneflowers, both of which had ironically been selected for their deer resistance.

Sheer tenacity (my husband called it stubbornness) kept me going. I was determined to find solutions rather than throw up my hands in frustration. We learned to protect young trees with temporary livestock fencing and took note of which plants the deer usually ignored, or only minimally damaged, such as barberries, spireas, and weigelas, and planted more of them, seeking out varieties in other colors, sizes, and shapes for added interest. As for the rabbits, I thwarted those by selecting fast-growing, taller grasses such as miscanthus (*Miscanthus sinensis*) and switchgrass (*Panicum virgatum*) over the shorter, juicier, and apparently tastier sedges (*Carex*) and Japanese forest grass (*Hakonechloa macra*). I also found that by waiting until spring to cut these back, the winter-dried stubble protected the tender new blades from their inquisitive noses. I refused to be beaten.

By 2015 the garden was finally beginning to look established. Then the voles moved in. Unlike their insect-eating cousin the mole, voles are vegetarian and have a voracious appetite for roots and bulbs. Plant lists couldn't help me this time—they ate anything and everything, and I began to lose mature trees as well as smaller shrubs and perennials to these pests. I researched various remedies, and we even adopted four barn cats, assured they would help. They didn't. I swear our lot were vegetarian pacifists, doing little to keep the rodent population down, preferring to stretch out in a quiet, sunny spot and snooze the day away. In desperation, I even tried a Molecat, an eco-friendly device that kills burrowing rodents with a percussive blast; however, after one unfortunate mishap due to user error that left my ears ringing for hours, I still wasn't any closer to a solution, and a large swath of yarrow (*Achillea millefolium* 'Paprika'), usually a highlight of the summer garden, was on the verge of complete collapse. At this time, a gardening friend offered a tip that voles seem to be repelled by members of the spurge family (Euphorbiaceae). I dashed out, bought all the donkey tail spurge (*Euphorbia myrsinites*) I could find at the nursery, and surrounded the yarrow with them. Sure enough the voles kept their distance, their exploratory tunnels stopping 6 inches or so away from the enclave. Success! While the voles and I continue to draw new battle lines daily, I have begun to regain control of the garden through the judicious use of assorted spurges, daily tunnel-stomping, and sheer force of will.

Today several discrete garden rooms offer a variety of experiences, from a welcoming front garden in shades of silver and blue to a bold, showcase island border visible from multiple vantage points. The critter-proof vegetable garden is only steps away from a newly installed patio, providing an abundance of organically grown produce—mere minutes from garden to table. The patio works equally well for a party of twelve or two, drawing Andy and me out on mild evenings to watch the sun go down over a glass of wine.

Each distinct garden thread is woven together in a carefully designed tapestry. Key colors, shapes, and textures are repeated, creating a sense of unity and rhythm, while focal points help to anchor vignettes, strengthen sightlines, and ensure year-round interest. The garden has become an extension of our home as well as my business, hosting our daughter's wedding, gardening events and workshops, magazine photo shoots, and even serving as the set for filming online video classes. The garden will never be perfect; something will always die, get eaten, or succumb to disease. But it's our little piece of heaven on earth and for now at least the wildlife seems to have agreed upon a truce.

Island Border

As soon as I saw the sun-bleached bigleaf maple (*Acer macrophyllum*), I knew I had stumbled across the perfect focal point for a showstopping "Karen lives here" border. I drew inspiration from the Northwest Perennial Alliance Border at the Bellevue Botanical Garden, which features meandering paths that immerse visitors in the sights, sounds, smells, and textures of its abundant plantings; however, in lieu of a typical high-maintenance perennial garden, I wanted to focus on trees and shrubs, creating a rich tapestry of colorful foliage accented by just a few choice perennials, selected for their good manners and easy-going attitude. Another consideration: as with most homes in the area, our water comes from a private well. Far from being a reliable resource, this has the very real potential to run dry after prolonged summer droughts; therefore, it was important to select plants that would be drought tolerant once established.

With the weeds removed and French drains added, we were ready for the fun to begin. Renting oversized Tonka toys to move boulders and more than a hundred yards of topsoil saved time, energy, money, and possibly our marriage. Andy became quite the expert at maneuvering three-man rocks with a utility Bobcat, nudging them gently into place. Likewise, a skidsteer was key to moving an existing cedar-shingled cabin to a better location. I knew it would perfectly anchor one end of the new border, while visually balancing the upright form of the maple and creating a charming focal point, visible upon entering the property as well as being perfectly framed by a new picture window in the kitchen. Three men, a big machine, heavy-duty straps, and some choice words later, the cabin was where we wanted it, although the grass looked as though we had hosted the Nascar races for the weekend.

Andy and I designed and built a triple arbor, paying special attention to scale. A single archway would have appeared diminutive in such a wide-open space, yet we didn't want a solid structure that would block views of the garden. An understated post and rebar design, each arch linked by a swag of marine rope, did the trick, giving dimension while maintaining transparency.

TOP LEFT Maneuvering large boulders several hundred feet across a bumpy pasture was one thing. Doing so without losing the moss was another.

LEFT With the porch roof braced and the deck temporarily removed, the cabin was ready to be moved to its new location.

BELOW The arbor not only highlights the pathway leading into the island border, it also serves as a year-round focal point, especially dramatic in the skeletal winter garden.

With the cabin, boulders, arbor, and a simple path in place we began to plant evergreen and deciduous trees that provided height and color while giving the illusion of transitioning into the forest beyond when seen from a distance. With all my plant choices, I first considered what design attributes I was looking for (e.g., a fast-growing deciduous tree with golden foliage for full sun), then researched options both in my personal library of garden books as well as online, before finally cross-checking my shortlist against the anticipated level of deer resistance. In this instance, I selected golden locusts (*Robinia pseudoacacia* 'Frisia'), which in this area do not sucker and have proven reliably deer-resistant thanks to their barbed branches. Other deciduous trees in this border include weeping willowleaf pear (*Pyrus salicifolia* 'Pendula'), paperbark maple (*Acer griseum*), Himalayan birch (*Betula utilis* var. *jacquemontii*), and Persian ironwood (*Parrotia persica* 'Ruby Vase'). The latter has since become a firm favorite with its spidery red winter flowers, interesting bark, and stunning range of foliage color, from green, gold, and orange to deepest purple and burgundy; it's also more of an upright vase shape than the broad-spreading species. A variety of conifers contribute year-round color, from the rich green of Hinoki cypress (*Chamaecyparis obtusa*) through the blue tones of a deodar cedar (*Cedrus deodara* 'Feelin' Blue') to the gold of an oriental spruce (*Picea orientalis* 'Skylands'). A Port Orford cedar (*C. lawsoniana* 'Wissel's Saguaro') is a columnar sentry, its cactus-like arms just begging to be hugged, while a white pine (*Pinus strobus* 'Blue Shag') offers a mound of teddy-bear softness. Color, texture, and personality—my sort of garden.

OPPOSITE In winter, when the fleeting summer orange of the perennial crocosmias dies down, the colorful peeling bark of this paperbark maple (*Acer griseum*) will be clearly visible.

BELOW An informal flagstone path, softened with creeping thyme (*Thymus serpyllum* 'Elfin') and flanked by billowing tickseed (*Coreopsis verticillata* 'Zagreb') now leads to the cabin. The pale gray-blue of the door echoes the color of nearby foliage.

From there I layered in shrubs that offer multi-season interest, skipping those with plain green leaves, even if they also contributed summer flowers, in favor of those that offered a framework of colorful foliage. I opted for a sunset-inspired color scheme—warm shades of gold, orange, red, and even hot pink accented by purple and softened with blue-greens. These vibrant shades would be sure to make a statement whether viewed up close or from a great distance. Plus, it was fun, so why not? We managed to salvage several mature shrubs from the original front garden, including a treasured Exbury azalea with highly fragrant blooms the color of liquid gold, lichen-encrusted branches, and jewel-toned fall colors.

LEFT At least he used the path.
BELOW Golden Dutch Master and white Mount Hood daffodils are highlights of the early spring garden.

After noting which path the deer took through this area, we planted a barrier of tall, thorny shrubs, including barberries (*Berberis thunbergii* 'Rose Glow'), hollies (*Ilex* 'Red Beauty'), and false hollies (*Osmanthus heterophyllus* 'Goshiki'). This spiky barricade forced the deer to either stay on the path or find another way around, protecting more delicate plants from their careless hooves.

Yet this border needed to be more than just a collection of deer- and drought-tolerant plants, albeit combined with an eye for nuances of color, contrasting textures, and intriguing forms. It also needed to be appealing in every season. Today, hundreds of yellow and white daffodils (*Narcissus*) herald the arrival of spring, flanking paths and clustering at the base of mossy boulders.

By early summer spirea foliage glows, golden leaves highlighting the abundant pink flowers. I have the last laugh even if the deer decide to nibble some of these blooms, as the new growth is a rich coral-orange, which I love even more than the flowers. Blue-flowering catmint (*Nepeta*) and geraniums (*Geranium* 'Rozanne') and glaucous-leaved smokebushes (*Cotinus coggygria* 'Old Fashioned') help to temper the fiery barberries and crocosmia, while a burgundy Japanese maple (*Acer palmatum* 'Fireglow') and a purple smokebush strike a high contrast

TOP Although deer pass through this border daily, the damage done is minimal. While the newly emerging foliage of dozens of orange daylilies (*Hemerocallis* 'Flasher') are tempting in early spring, a quick spritz with a proprietary deer repellent seems to be enough to persuade them to do their taste-testing elsewhere.

ABOVE Strategically placed Arkansas blue star (*Amsonia hubrichtii*) is an autumnal star. The wispy foliage shifts through shades of gold, orange, burgundy, and purple, each kaleidoscopic variation more camera-worthy than the last when set against a purple smokebush.

against neighboring golden conifers and a sweep of incendiary orange daylilies.

The cabin is highlighted by a late-summer meadow of golden black-eyed Susans (*Rudbeckia fulgida* var. *sullivantii* 'Goldsturm'), the rich brown central cone of each daisy echoing the dark cedar shingles. As summer moves into fall I find myself frequently reaching for my camera to photograph the feathery Arkansas blue star (*Amsonia hubrichtii*) as it transitions through shades of copper and gold.

Meanwhile a katsura (*Cercidiphyllum japonicum*) exudes its distinctive scent of candy apples as the entire border embarks upon an autumnal display that lasts from September to late November. Even in winter this border is noteworthy, the bare tree branches and colorful bark, seed heads, grasses, and winter-hardy leaves etched with hoarfrost, highlighting their patterns and silhouettes. Like the rest of the garden this border is constantly evolving, but it has come a long way from its inauspicious beginnings.

Woodland Border

"Keep going!" I shouted, waving the driver of the excavator onward as he continued to carve out a drainage trench that led away from the island border toward a scruffy stand of cottonwood and alder saplings on the edge of the property. "Just wiggle a bit," I added, hoping he would understand that I was asking for a natural-looking streambed rather than requesting he break out into some exotic dancing as he drove. Miraculously, he did, and after some more wiggling, we discovered water bubbling up out of the ground along this newly excavated trench, which explained in part why this previously undeveloped area was always so wet.

Until that moment I hadn't given much thought to this part of the garden but suddenly I could envision a seasonal stream flanked by groves of deciduous trees, their canopies filtering the strong overhead sunshine to provide the perfect dappled conditions needed by the many shade-loving plants I had brought from our previous garden. This area would become a somewhat secluded, peaceful strolling garden that would provide respite from the intense summer sun. But first we had to remove countless beer cans, roadfill from a nearby construction site, and of course more weeds.

To prepare the ground for planting I had to tackle the reed canarygrass (*Phalaris arundinacea*), a highly invasive noxious weed I now know to be an indicator of very wet land. Spraying with herbicides was out of the question: there was a danger of contaminating the salmon spawning streams into which this newfound water would ultimately flow. Instead I layered cardboard, 4–6 inches of compost, and arborist chips, creating a weed-smothering sheet mulch that broke down over the course of a year, leaving in its wake a workable soil. A few pernicious clumps of the aggressive grass persisted, but I eventually won the battle. To make year-round access possible, Andy built two wooden bridges to traverse the stream, and I added a narrow circular path of wood chips.

Three ornamental pear trees that backed the original border were kept, as they offered height, maturity, and a sense of boundary between what was visible from the main garden and this new, private space. I supplemented them with river birch (*Betula nigra* 'Heritage'), Himalayan birch, red maples (*Acer rubrum*), and columnar Swedish aspens (*Populus tremula* 'Erecta') for their seasonal water tolerance and bark interest, together with several golden locusts and a few smaller Japanese maples in the drier areas. This part of the garden is extremely wet during the winter, not only from the stream and perched water table but also from a pond on neighboring land that overflows into our garden. The extreme contrast between winter wet and summer dry made selecting plants for the woodland border very challenging. In addition, this area cannot be reached with a hose, and without a water source nearby, even hand watering wasn't practical.

My aim was therefore to select plants that were appropriate aesthetically for a woodland yet were also able to cope with these less-than-perfect conditions as well as being both rabbit- and deer-resistant. I noticed that the deer seemed to have a regular route through the woodland, as some areas were

INVASIVE PLANTS

Some of the plants featured in this book may be invasive in your area—please consult your local Cooperative Extension office or check the USDA website (plants.usda.gov) before planting.

browsed more heavily than others. I took advantage of this by placing plants that the deer seemed to reliably ignore—drooping fetterbush (*Leucothoe fontanesiana* 'Rainbow'), hellebores (*Helleborus*), autumn ferns (*Dryopteris erythrosora*)—closest to their route and kept the temptations of my big, juicy hostas farther away. Having said that, the slugs can turn those hostas to lace in a matter of hours. There's always something.

My quest for tough plants that met all these criteria has had mixed results. Oregon grape (*Mahonia*) hates wet winter soils yet seems to revive in spring. Twig dogwoods (*Cornus*) are not a shrub I would recommend for deer-prone gardens, as the new growth is always nibbled in spring, but since these shrubs thrive in wet soil and their winter color is so striking I couldn't resist trying them. I planted mine under some tall red maples, fronted by low-growing sedges and ferns, none of which are of interest to passing

RIGHT The streambed affords the moisture-loving Rodgers' flower (*Rodgersia podophylla* 'Rotlaub') ideal growing conditions. Its leaves are a rich bronze as they emerge in spring, finally turning green.

BELOW Two simple wooden bridges were constructed to span the seasonal stream that runs through the woodland.

deer, and so far they have gone unnoticed. Virginia sweetspires (*Itea virginica* 'Little Henry') are fabulous semi-evergreen shrubs in these challenging conditions; they thrive in areas with the wettest soil, which seems to hold enough moisture to get them through most summers, although they may defoliate in especially dry years. Their spring flowers may get nibbled but not enough (usually) to spoil the show, and the rosy color of the fall foliage alone is reason enough to add them to any garden. Astilbes on the other hand have not been successful—in summer the soil dries out too much for these moisture-loving perennials.

The woodland border looks its best in spring when the cobalt-blue flowers of Catlin's Giant bugleweed rise on short fat spikes above glossy black basal leaves; these line the streambank, amplifying the river effect. Bringing a little horticultural sunlight to the scene are several snowberries (*Symphoricarpos* ×*chenaultii* 'Blade of Sun') growing on either side of the bank; the small, bright yellow leaves on this ground-hugging layering shrub are truly delightful at this time of year. Elsewhere in the border, several varieties of bleeding heart (*Lamprocapnos spectabilis*) have their moment of glory together with some English primroses (*Primula*

OPPOSITE Layering creates an interesting woodland walk, the shrubs and perennials forming an understory to taller trees; bugleweed (*Ajuga reptans* 'Catlin's Giant'), bearberry cotoneaster (*Cotoneaster dammeri*), and other groundcovers control soil erosion on the streambanks.

ABOVE Front to back: barberry (*Berberis thunbergii* 'Red Carpet'), snowberry (*Symphoricarpos ×chenaultii* 'Blade of Sun'), and bugleweed (*Ajuga reptans* 'Catlin's Giant') mingle to create a colorful tapestry on the streambanks.

RIGHT A cluster of primroses (*Primula* 'Dorothy') and a garden memento from our daughter's childhood, waiting to be discovered by the next generation.

vulgaris) and cowslips (*P. veris*) that I have grown from seed.

To truly appreciate the magical qualities of the woodland garden, one needs to explore through a child's eyes. Diminutive stone statues and soft, mossy rocks are waiting to be discovered; fragrant English bluebells (*Hyacinthoides non-scripta*) become fairy hats in imaginative hands; posies of petite blooms can be picked for the table; tightly furled fiddleheads could be a tasty treat; the bridges invite a time-honored game of Pooh sticks with a friend. For those a little older but still children at heart, a stone bench invites quiet contemplation, while a whimsical collection of Three Blind Mice is sure to draw a smile. This is an enchanting place for those that take the time to look and to listen.

Front Garden

When we first viewed the house in the summer of 2009 a rampant wisteria dominated the front garden, all but obscuring the path, while the few shrubs and abundant perennials were completely engulfed by the attractive but highly invasive bishop's weed (*Aegopodium podagraria* 'Variegatum') and a lifetime supply of mint. I knew straight away that virtually everything would have to go. Even though the plants were deer-resistant and mostly tolerant of drought, they were going to need far more time to manage than I was willing to invest, and this was also very much a summer-focused rather than a year-round design. Even the overall structure of the space wasn't safe. As the position of the front door had been altered during the house remodel, the original path no longer lined up correctly. Likewise, when we converted a small garage to living space, the driveway design needed to be reconfigured.

Our vision was to create lush planting beds that would frame the home much as a traditional foundation planting plan might do, but additionally we decided to incorporate a small sitting area in the front garden and sweep the path in a much wider arc to make the approach to the front door feel more of a journey. A focal point was needed that would draw the eye and feet toward the home's entrance. We used two sleek turquoise containers, converting the smaller one into a bubbling fountain and adding a stunning glass sculpture by Seattle artist Jesse Kelly to the larger vessel. Together they made the perfect statement, visible upon entering the property, when walking to the front door, or when viewing the garden from inside the home. They also helped to establish the color scheme for the front garden.

While both the woodland and island borders can be seen, at least in part, from the front garden, their physical distance makes it possible to have a different color palette in each area without the effect appearing too busy. The front garden is primarily blue, silver, and white with just a few accents of purple and chartreuse. These two accent colors can be found in the glass art as well as in all the other garden borders, so there is a subtle connection even as this space asserts its own identity.

LEFT The color palette for the front garden was inspired by the dramatic glass sculpture and containers.

BELOW Sea holly (*Eryngium* 'Sapphire Blue') erupts from a carpet of fragrant sweet alyssum (*Lobularia maritima*). Both are remarkably tolerant of drought, as well as being both rabbit- and deer-resistant.

Backlit by the sun, the thread-like foliage of a Japanese maple (*Acer palmatum* 'Koto-no-ito') appears to glow.

It took almost two years to dig out every last piece of those aggressive weeds, but it was worth the time and effort. Now the summer display includes mounding barberries (*Berberis thunbergii* 'Concorde'), their rich purple foliage enhancing the superbly floriferous blue Rozanne geraniums and delicate wands of gaura (*Gaura lindheimeri* 'Whirling Butterflies') that appear to dance in the slightest breeze from late spring through fall. All are backed by Magical Fantasy weigela (*Weigela florida* 'Magical Fantasy'), a delightful cultivar with crisp white and green variegated foliage, as well as other five-star flowering shrubs including a golden-leaved dwarf bluebeard (*Caryopteris* ×*clandonensis* 'Lil' Miss Sunshine') and a beautiful variegated rose of Sharon with ruffled lavender blooms (*Hibiscus syriacus* 'Summer Ruffle'). I often add sweet alyssum (*Lobularia maritima*) to edge the paths and borders, its honey-like scent perfuming the air. As well as being deer-resistant and remarkably drought tolerant, the flowers attract many different pollinators, their gentle humming adding to the sounds of summer.

Although the front garden display starts to wind down in fall, one highlight is a specimen Koto-no-ito Japanese maple (*Acer palmatum* 'Koto-no-ito'). The name translates to "harp strings," appropriate for the delicate, thread-like leaves that turn astounding shades of gold and copper as autumn progresses.

A variegated Canadian hemlock (*Tsuga canadensis* 'Gentsch White'), pruned each spring to promote the brightest color, junipers (*Juniperus squamata* 'Blue Star'), and dwarf pines ensure plenty of winter interest, while the pristine white bark of a Himalayan birch creates drama, especially

The vegetable garden features a double fence to keep deer out. To date, the tactic has worked, and we have yet to discover any confused deer wandering within it.

when uplit at night. It seems no time at all before the garden awakens again in spring with daffodils, English bluebells, purple ornamental onions (*Allium*), luscious peonies, and the heady perfume of old-fashioned Cheddar pinks (*Dianthus gratianopolitanus* 'Firewitch').

While deer do investigate this garden, their visits are primarily confined to winter months when they do little damage other than standing on emerging bulbs. I have even successfully grown two roses in this area, although I do spritz them with repellent when in full bud as a precaution. Regardless, this newly designed space welcomes friends and family, offering an intimate setting in which to sit and listen to the sounds of the garden—the gentle bubbling of a fountain, the distinctive whir of a hummingbird's wings, and the lazy droning of hundreds of honey bees enjoying the garden as much as we do.

Vegetable Garden

It began with the War of the Parsnips. Our family loves these roasted root vegetables for Thanksgiving dinner as well as in hearty winter soups and casseroles, so I was excited that our new home had an existing fenced area suitable for growing them. So were the voles. They completely decimated my crop that first year, eating every single delicious root and leaving me only the leafy tops. I was *not* impressed. Thus began the War of the Parsnips and a complete redesign of our vegetable garden in order to keep out deer and rabbits as well as those pesky burrowing rodents. However, being close to the house and highly visible, the structure needed to be attractive as well as functional.

At 40 feet by 45 feet, the new plot exceeded the square footage of the house, so surrounding it with a traditional 8- to 10-foot-tall deer fence would have given the impression of a fortress and completely overwhelmed our single-story home. We opted instead for two 5-foot-high fences spaced 5 feet apart, with 8-foot-tall pergolas, planted with vines, at the two entrance gates. The principle is that deer won't jump when they can't see what is on the other side, and although they could easily clear one fence, they can't land safely between the two, or jump both at the same time. We used panels of galvanized hog wire with graduated openings for the fencing; the smaller size at the base keeps the rabbits out, and the deer can't get their muzzles through the larger gaps at the top. Conveniently, the panels also serve as ideal structures on which to grow sweet peas or as a framework for espaliered fruit trees.

That took care of the rabbits and deer, but we still had the troublesome voles to deal with. We sunk quarter-inch galvanized mesh hardware cloth 3 feet deep around the perimeter of the vegetable garden, attaching it to 10-inch-high boards installed at the base of each fence section. Voles can't get through such small mesh, and we didn't think they could scale the vertical 10-inch boards—but sadly we were wrong on the latter account. If we were to do this again, we would use 18-inch vertical base boards. However, a combination of 12-inch-high raised beds for the vegetables together with entry being harder to gain for all but the most athletic of those sneaky little pests means we have enjoyed lots of parsnips ever since, and the voles have gone hungry.

I have to confess that none of this was cheap. You might well say, "Karen, there are perfectly good stores where you can buy parsnips." I know. But where's the fun in that?

Top 10 Plants

WHIRLING BUTTERFLIES GAURA (*Gaura lindheimeri* 'Whirling Butterflies'). Delicate wands of pink-suffused white flowers dance above the foliage from late spring through fall. In colder areas this drought-tolerant, deer-resistant perennial may die back to the ground in winter, so avoid cutting back old stems until spring. In warmer climates, it may remain a woody shrub and self-seed. Grows 3 feet tall and wide. Full sun, zones 5–9. Deer resistance: A.

PAPRIKA YARROW (*Achillea millefolium* 'Paprika'). Feathery, aromatic gray-green foliage is the perfect foil for the abundant summer blooms that open bright crimson and transition through shades of pink and terracotta before fading to tan. Deadheading at this stage will encourage a second wave of flowers. Reliably drought tolerant. Grows 2 feet tall, 3 feet wide. Cut this herbaceous perennial back in late fall or early spring. Full sun, zones 3–9. Deer resistance: B.

BLUE STAR JUNIPER (*Juniperus squamata* 'Blue Star').
A low-growing blue-gray conifer, 2 feet tall, 3–4 feet wide, with short, stiff branches and spiky needles. Ideal for containers or the landscape. Full sun, zones 4–8. Deer resistance: B.

LIME GLOW BARBERRY (*Berberis thunbergii* 'Lime Glow').
A thorny, deciduous shrub, 4 feet tall and wide, with marbled lime-green and creamy white foliage that turns vivid shades of crimson in fall. Prune by a third in spring to encourage the colorful new growth. Full sun, zones 4–8. Deer resistance: A.

BLADE OF SUN SNOWBERRY (*Symphoricarpos* ×*chenaultii* 'Blade of Sun').
An outstanding deciduous groundcover for a woodland garden with wide-spreading branches that form dense layers of golden foliage. Emerging brilliant yellow, the color fades to chartreuse in deeper shade, returning to gold in fall, when small pink berries add to the scene. Grows 1–2 feet tall, 3 feet wide. Partial shade, zones 4–9. Deer resistance: B.

RUBY VASE PERSIAN IRONWOOD (*Parrotia persica* 'Ruby Vase').
Interesting bark, winter flowers, multi-toned foliage, and stunning fall color—this slender tree deserves a place in every full sun–partial shade garden, eventually reaching 30 feet tall, 12 feet wide. Emerging leaves on lower branches may be nibbled in spring. Zones 5–7. Deer resistance: B.

DOUBLE PLAY GOLD SPIREA (*Spiraea japonica* 'Double Play Gold').
Emerging spring foliage is copper, maturing to gold in summer, when the entire shrub is covered with hot pink flowers that attract bees, butterflies, and the occasional deer. The floral display is so abundant that the few deer-pruned blooms won't be missed, and the new flush of foliage color more than makes up for the minor damage. Grows 3–4 feet tall and wide. Full sun–partial sun, zones 4–9. Deer resistance: B.

ARKANSAS BLUE STAR (*Amsonia hubrichtii*).
Early spring flowers are blue, but this perennial is all about the feathery green foliage, which turns shades of orange in fall. Grows 2 feet tall and wide; extremely drought tolerant. Full sun, zones 5–8. Deer resistance: A.

ENGLISH BLUEBELLS (*Hyacinthoides non-scripta*).
Unlike their invasive Spanish cousins, English bluebells are well behaved and do not spread aggressively but rather naturalize slowly, forming carpets of fragrant spring blooms, 1.5 feet tall. Plant bulbs in fall. Full sun–partial shade, zones 5–8. Deer resistance: B.

SAPPHIRE BLUE SEA HOLLY (*Eryngium* 'Sapphire Blue').
One of the bluest sea hollies, the spiky bracts of this herbaceous perennial make a bold statement in the summer garden. Grows 2.5 feet tall, 1.5 feet wide. A favorite of flower arrangers, bees, and butterflies but mercifully completely ignored by deer and rabbits in my garden, although Rutgers lists the genus as being frequently severely damaged. Full sun, zones 4–8. Deer resistance: D.

A Country Garden
where daunting challenges become exciting opportunities

It took more than a hurricane, flooding, and hungry deer to squash the Blooms' vision of an idyllic country garden, yet these were just some of the challenges they faced in their new home.

Minutes from New York yet a world away, Essex Fells is a small borough of winding, tree-lined roads and family homes set well back behind expansive, manicured lawns. The ambience suggests a bygone era when life moved at a slower pace. Many of the residents either never leave or return to raise their own families, just as Ruth and William Bloom chose to do. With siblings still in the neighborhood, the Blooms were thrilled to have the opportunity to purchase a 1910 Colonial-style home for their family, expanding its original footprint to better suit their lifestyle and softening the architecture with the addition of stone detailing.

QUICK FACTS
LOCATION: Essex Fells, New Jersey (zone 6)
SOIL TYPE: sandy loam
PROPERTY SIZE: 2 acres
PROBLEM CRITTERS: deer and two rambunctious dogs
OTHER CHALLENGES: steep hillside; poor drainage

DESIGN CREDITS:
home of William and Ruth Bloom
designed by Susan Cohan, Susan Cohan Gardens

Challenges and Dreams

With the home's renovations complete and her three boys and dogs in need of play space, Ruth turned her attention to the outside. For such a large lot, there was surprisingly little garden or usable outdoor living space. A narrow brick patio and lap pool gave way to a weedy hillside culminating in a shallow depression filled with Norway maples. Also at the foot of the hill was an additional single-story dwelling, formerly used as a rental property but now a hindrance. William, a volunteer firefighter at the time, had an inventive solution for its removal: he invited the local fire department to use it for a training drill.

With the structure removed, Ruth hired architect Francis C. Klein to design a barn-style pool house with a full catering kitchen together with a pool and spa, a multi-functional area that would be suitable for large-scale entertaining as well as more intimate gatherings. Ruth also began clipping magazine articles that encapsulated her vision for a garden around the pool, and it was at this point she contacted landscape designer Susan Cohan, an early collaboration that turned out to be crucial to the success of the overall project. Known for her functional designs and thoughtful spatial arrangements, Susan relocated the spa for better flow and

RIGHT Spirea (*Spiraea japonica* 'Little Princess') tumbles onto a new stone wall adjoining the home. All plantings behind it are beyond the reach of all but the most tenacious of deer.

BELOW Originally constructed with aged brick and ornate columns, the Bloom family updated and expanded the 1910 Colonial-style home to give it a more welcoming, country feel.

With a fully equipped barn-style pool house, multiple seating options, and a portable firepit, the pool area has become the heart of the garden with plenty of room for relaxing or for entertaining.

designed a shaded pergola, which the architect then detailed and built. The final design of the pool area maximized the space for dining, swimming, and entertaining. At first hesitant to embark on such a major project, the Blooms now say this is one of the best decisions they ever made. They open the pool in early April and don't close it until Halloween. "The dogs love it," Ruth laughs, "and I love looking down on it. It's so pretty."

Unfortunately, construction of the pool area compounded what was already a major challenge: poor drainage exacerbated by a high water table. After snowmelt or major rainstorms, runoff from the hill and the slightly elevated road that runs behind their property regularly flooded the lower lawn. Now it would also receive runoff from the new, impermeable patio and overflow from the pool itself. Susan worked with a civil engineer to design an underground system of eighteen gravel-filled pipes installed at the base of the hill to allow any standing water to percolate. The system is designed such that any overflow will drain into a new dry stream, but to date that has never been necessary.

Also high on the wish list was the need for a large lawn for the three boys to kick a ball about and for playful dogs to run freely. For herself, Ruth wanted somewhere to grow cut flowers, and William requested a better flow from the home's terrace down

An underground drainage system ensures the large, level lawn at the base of the hillside will never flood again.

to the new pool area, currently navigable only by a steep grassy slope.

And the challenges didn't end there. Deer visit this garden daily, so the plant selection needed to be deer-resistant since everything outside the fenced pool area is open to them. Ruth has noticed that when the dogs are present the deer don't seem to be quite as brazen and believes that sprinkling dog hair around the property's perimeter may be something of a deterrent. Finally, although this is their primary home, the family also has properties in Boston and on the Jersey Shore, so Ruth wanted the garden to look its best in spring and fall since they are often away during peak summer weeks and in winter.

Dreams Realized

As an experienced designer, Susan was more than equal to these formidable challenges and, after mitigating the issues of standing water, began by creating the planting beds that surround the pool area. Using dry-stacked bluestone, superior in quality and thickness to paver stone, she fashioned raised planters that complemented the gray tones in the stone veneer used on both the home and pool house. By elevating the plantings, the designer ensured that their roots would never be in saturated soil. Loose spheres of boxwood (Buxus 'Green Velvet') serve as a structural evergreen backdrop to an informal medley of shrubs and perennials in cool shades of pink, lavender, blue, and white—very much in keeping with Susan's design aesthetic: "Provide strong structure, then plant with abandon." Ruth was in complete agreement; she too prefers this low-key, harmonious palette over bolder shades, restraint rather than excessive variety.

Peonies remind Ruth of their wedding day, so she selected several varieties to include in her new garden. Their fragrant, blowsy appearance brings her much joy, yet she hesitates to cut them for the home for fear of spoiling the display. The solution? She is currently planning a dedicated peony cutting garden elsewhere on the property.

TOP Capped, dry-stacked bluestone walls define the raised planting beds that surround the new pool house.

ABOVE Two weathered tuteurs anchor the raised planting beds, adding height, year-round structure, and a sense of history to the space.

LEFT "I love them all!" exclaims Ruth when asked to name her favorite peony. Sarah Bernhardt (*Paeonia lactiflora* 'Sarah Bernhardt'), a popular heirloom variety, is one of many in her garden.

A Country Garden

Taming the Hillside

Conquering the steep hillside took heavy-duty equipment and several tons of additional rock. Terraces were crafted by supplementing existing boulders found on site and arranging them as a series of rocky outcroppings, then backfilling the deep planting pockets with a well-draining soil blend. The newly terraced areas look as though they have always been part of the landscape.

The casual planting within these terraces may appear unintentional, yet Susan carefully orchestrated each combination and individually placed every plant. Trained at the English Gardening School, Chelsea, she was taught to repeat either a color or a plant on a diagonal axis. Susan has adapted this technique slightly, feeling that contrary to popular opinion, *even* numbers of plants rather than odd make for better drifts, which she then interlocks along the length of the border into a textural ribbon. The effect is magical.

While the uppermost tier is anchored with larger shrubs, including lilacs and doublefile viburnum

LEFT The picture of innocence, Remy pauses between toy-chasing antics. The designer avoided brittle or delicate plants in favor of those that could withstand the fun-loving dogs.

BELOW To make the slope more navigable on foot, a series of reclaimed granite curbs were set into the grade, linking the upper terrace to the new pool area.

OPPOSITE Undulating clumps of ornamental grass (*Miscanthus sinensis* 'Gracillimus') weave between violet-blue spires of sage (*Salvia nemorosa* 'Caradonna') and other sun-loving perennials, shafts of evening light casting an ethereal glow.

(*Viburnum plicatum* f. *tomentosum*), both levels are densely planted with drifts of perennials and tall drifts of miscanthus and other grasses that weave together in an informal manner. The sturdy grasses hold up especially well to the dogs' escapades, and all the plants are safe for their occasional foraging. Another unexpected issue arose when one of the Blooms' black Labrador retrievers, Remy, began to obsessively chase butterflies. Susan therefore tried to select plants that were not butterfly magnets and abandoned some earlier choices that proved problematic in that regard. To allow the plants on the hillside to become established, the Blooms erected temporary fencing to keep the dogs out, admitting this proved difficult since the area had originally been their playground. The fencing worked well, however, until the dogs discovered that they could walk along the stone wall and jump into the beds. "You'd think with all the lawn and the swimming pool they'd have enough!" laughs Ruth.

Although Ruth spearheaded the overall landscaping project, William requested an abundance of blue flowers in the garden, both because that is his favorite color and because it plays off the gray tones in the stone so effectively. To this end Susan included, among others, Arkansas blue star (*Amsonia hubrichtii*), false indigo (*Baptisia australis*), catmint (*Nepeta racemosa* 'Walker's Low'), and several varieties of sage, guaranteeing blues are present from spring to fall. To satisfy Ruth's desires, more pink flowers have since been added, including betony (*Stachys officinalis* 'Hummelo') and a dwarf beebalm (*Monarda* 'Petite Delight'), while accents of white and yellow are sprinkled across the green backdrop.

Creative Solutions

Bordering the lawn, a layered planting of evergreen and deciduous trees and shrubs provides privacy, buffering the property from neighbors and the road; included are spruce (*Picea*), river birch

The dry stream is both functional and decorative, winding between sycamore trees and edged with moisture-loving perennials.

(*Betula nigra* 'Heritage'), southern magnolia (*Magnolia grandiflora* 'Little Gem'), and Japanese tree lilac (*Syringa reticulata*). At the southwestern boundary, a natural-looking dry stream introduces an eye-catching detail as it meanders past sycamores (*Platanus occidentalis*), flanked by astilbes, irises, and ferns, all of which thrive in the moist, partially shaded setting. A granite bridge completes the scene. Susan laughs as she explains how the crew built the dry creek bed: "I'd bring beer and we'd all sit on the grass and drink. Occasionally we'd get up and place a boulder or two—then drink some more beer. The result is a very naturalistic design!"

Not everything has gone according to plan, however. "There have been some spectacular failures," confesses Susan, recalling how she doggedly

LEFT A granite bridge spans the dry stream, nestled between strategically placed boulders that were handpicked by the designer.

BELOW Bluestone pavers are laid to resemble a giant checkerboard; a trellis affixed to the stone wall repeats the grid.

replanted hellebores (*Helleborus*) and hay-scented ferns (*Dennstaedtia punctilobula*) three times in one problematic area. She's still not certain what caused their demise. And Rozanne geranium (*Geranium* 'Rozanne'), formerly ignored by the deer, has suddenly been eaten by either the deer or possibly rabbits. Ruth admits she has been advised to spray deer repellent on vulnerable plants but has a fairly relaxed attitude to any minor damage, preferring to wait and see how bad it gets rather than taking a preemptive approach.

On one unforgettable occasion, a sudden loss became an opportunity for creativity. In 2012 Hurricane Sandy devastated the East Coast, destroying homes and flooding streets, leaving a trail of debris in its wake. One of the casualties was a mature Japanese tree lilac on the upper level of the Bloom garden. A casual conversation with Susan sparked the idea for placing an outdoor checkerboard in that area, which Ruth immediately had installed. The square bluestone pavers make a distinctive patio that looks as though it has always been there.

This sense of place resonates throughout the garden, from the design of the pool house meticulously crafted to appear as though it were original to the home, to the expertly organized pool area with defined activity zones. Yet it goes deeper than that for Ruth, whose love of gardening was inspired by both parents, especially her mom, who loved her rose and tomato gardens. Like her mother, Ruth has discovered that the garden is her "happy place," where she can be mindful of her surroundings and briefly cast worries aside. After years of thoughtful planning and nurturing, she also has a profound sense of accomplishment watching the garden grow. "Just looking at it makes me happy," she says with a smile.

Top 10 Plants

JAPANESE TREE LILAC (*Syringa reticulata*). A deciduous tree, 20–30 feet tall, 15–25 feet wide. It blooms later than the shrub lilacs, producing large clusters of fragrant, creamy white flowers that attract butterflies and other pollinators. Full sun, zones 3–7. Deer resistance: B.

GREEN VELVET BOXWOOD (*Buxus* 'Green Velvet'). One of the most reliable varieties for the East Coast, with good resistance to bronzing. Grows 3–4 feet tall and wide. Best protected from desiccating winds. Full sun–partial shade, zones 5–8. Deer resistance: A.

LITTLE PRINCESS SPIREA (*Spiraea japonica* 'Little Princess'). A dense, mounding deciduous shrub that remains compact without pruning, 1.5–2.5 feet tall, 2–3 feet wide. Clusters of pink flowers cover the mint-green foliage from late spring to midsummer, with occasional repeat bloom thereafter. Full sun, zones 4–8. Deer resistance: B.

MOONSHINE YARROW (*Achillea* 'Moonshine'). Best grown in lean, dry soils and full sun, this perennial features gray foliage and flat sprays of sunshine-yellow flowers in early summer. Deadhead to prolong blooming and divide clumps every three to four years to reinvigorate. Grows 1–2 feet tall, 1 foot wide. Zones 3–8. Deer resistance: A.

NIKKO SLENDER DEUTZIA (*Deutzia gracilis* 'Nikko'). A compact shrub, 1–2 feet tall, 2–5 feet wide, covered with tiny, fragrant bell-shaped blooms in late spring. The green foliage turns red in fall. Full sun–partial shade, zones 5–8. Deer resistance: B.

HUMMELO BETONY (*Stachys officinalis* 'Hummelo'). Resembling a pink-flowering sage, betony is noted for its late spring floral display. Drought tolerant once established, this creeping perennial, 1.5–2 feet tall and wide, can be used to naturalize sunny areas. Full sun, zones 4–8. Deer resistance: A.

MAY NIGHT SAGE (*Salvia ×sylvestris* 'Mainacht'). A popular perennial for dry, sunny areas, with deep blue flowers in early summer that attract bees and hummingbirds. Grows 1.5–2 feet tall, 1–1.5 feet wide. Remove spent flowers and provide supplemental water to encourage rebloom. Full sun, zones 4–8. Deer resistance: B.

PACHYSANDRA (*Pachysandra terminalis*). Often selected as a reliable groundcover for dry shade, pachysandra spreads slowly to form a dense carpet of evergreen rosettes, 0.5–1 foot tall and 1–1.5 feet wide. Tiny inconspicuous white flowers appear in early spring. Partial shade–full shade, zones 5–9. Deer resistance: A.

HARTLAGE WINE ALLSPICE (*Calycanthus ×raulstonii* 'Hartlage Wine'). Tolerant of a wide range of soils and full sun but thriving in partial shade, this multi-stemmed deciduous shrub is covered with maroon flowers that fade to wine-red. The sterile blooms have an unusual fruity fragrance. Grows 8–10 feet tall and wide. Zones 5–9. Deer resistance: B.

DARK TOWERS PENSTEMON (*Penstemon* 'Dark Towers'). A clump-forming perennial, 1.5–3 feet tall, 1–2 feet wide, with ruby-red foliage that holds its color well throughout the season, and pale pink flowers that appear on tall stems in early summer. The tubular flowers attract bees and hummingbirds. Full sun, zones 3–8. Deer resistance: C.

A Desert Oasis
where neither heat, nor drought, nor deer diminish the mirage

Appearing like a mirage in the suburban Texas landscaping desert, the Penick garden stands out for its very existence. Many of the neighborhood lots have been reduced to a token evergreen accenting a drought-stressed lawn, defeated by the harsh climate. Pam and David's northwest Austin garden, however, is a showcase of bold, architectural succulents and palms in shimmering shades of silvery blue, mingling with diaphanous grasses and choice native shrubs, all enhanced by an array of intriguing artistic details. Clearly, this is not your typical garden—but then Pam is not your typical designer.

Originally from South Carolina, Pam met her future husband David while at school in Houston, where she was studying English. David's career in software later took them to Austin, where Pam worked as an editor before becoming a full-time mom. It was during those years that

QUICK FACTS

LOCATION: Austin, Texas (zone 8)
SOIL TYPE: clay, limestone, and caliche
PROPERTY SIZE: 0.33 acre
PROBLEM CRITTERS: deer, armadillos, rabbits
OTHER CHALLENGES: extreme heat and drought; dry shade; flash floods; watering restrictions

DESIGN CREDITS:
home of David and Pam Penick
designed by Pam Penick

she discovered a passion for gardening and began her blog, *Digging*. That award-winning gardening blog was the gateway to many wonderful friendships as well as professional writing opportunities, but in those early days it helped to launch her career as a landscape designer, since friends and neighbors asked Pam to help them replace their thirsty lawns with drought-tolerant native plants.

Several years and two houses later, Pam and David began to look for a larger family home with good schools nearby for their growing children. Pam's initial impression of this 1970s ranch? "It was all roof!" While pleasant, the garden was equally uninspiring. Narrow planting beds bordered a series of small lawns, and the sloping ground appeared to swallow the house. Once inside, however, the Penicks found the home's layout eminently livable, and they loved the rear outdoor living spaces with a deck, patios, and pool—a necessity in the brutal Texas summers. Over time, and as budget allowed, Pam and David knew they could update the home and redesign the landscaping to better reflect their personalities, and so in 2008, this slice of central Texas became theirs.

It didn't take long for Pam to realize that the soil in this garden was very different from the deep black gumbo clay she was used to, instead consisting of clay, limestone, and caliche, a rock-like substance created when calcium carbonate cements soil particles together. While not a total surprise to Pam, having been warned by the previous homeowner that digging holes would be a major undertaking, it was nevertheless part of her gardening learning curve.

Contorted live oaks (*Quercus virginiana*) preside over a medley of drought-tolerant plants that thrive in the dappled shade.

ABOVE A young fawn resting in the shade of a tree appears to be totally unafraid of human visitors.

RIGHT Pam has a designer's eye for combining drought-tolerant natives. Here the spiky succulent foliage of pale-leaf yucca (*Yucca pallida*) mingles with fuzzy gray-woolly twintip (*Stemodia lanata*)—a study in contrasting textures.

A more obvious challenge was the five or six deer that visited daily, seeking a shaded spot for a quiet nap, apparently unperturbed by humans. Although Pam had never shared her previous gardens with deer, her experience as a designer had provided insights into the types of plants that were deer-resistant in Texas. Armadillos were also regular visitors, their terrible eyesight causing them to bump into things, much to the family's amusement. While not causing direct damage to the plants, they love freshly turned soil and will uproot newly planted bushes in search of earthworms and grubs.

Perhaps the most significant and unwelcome surprise occurred during a torrential rainstorm not long after they moved in. Since the front garden and driveway slope toward the house, water poured like an unchecked river toward it, destroying landscaping, washing away mulch, and flooding the pathway, making access to the front door impossible without rain boots. Clearly, that problem needed to be addressed promptly.

Designing a Mirage

Pam envisioned a strolling garden with several paths leading off the semicircular driveway, inviting exploration and offering a variety of garden experiences. Before level paths could be created, however, the slope needed to be addressed, so Pam designed a low limestone wall to terrace the front garden, creating a new path between the landscape and the house. Now the garden appears to frame the home rather than engulf it. To address the tough caliche soil, Pam brought in additional topsoil to create berms for increased planting depth. She also top-dressed new beds with a mixture of compost and decomposed granite, aiming to improve the depleted soil while facilitating drainage.

Several measures were employed to slow water down during a heavy storm, allowing time for it to percolate into the soil. A dry streambed was installed across the front of the house to divert water during flash floods, while stone baffles set in the mulch and wooden baffles in the gravel paths slow the water down.

Equally high on the priority list was removing a patch of lawn by the front door which proved difficult to mow, together with a narrow strip of shrubs that had been planted too close to the home's foundation. In its place, Pam pictured a gravel courtyard accented with a collection of sculptural planters. Unfortunately, the crew hired to remove the turf also regraded this confined space, exacerbating the existing drainage problem. With the next rainstorm, the Penicks realized that water was once again trying to enter their home rather than being diverted into the new dry streambed. A sump pump installed in the far corner of the gravel courtyard now siphons water away from the foundation, directing it into the dry streambed. Pam and David also installed gutters at the front of the home, with the downspouts moving rainwater out to the driveway and into the streambed. Finally, the drainage problem was solved.

The original front garden included several other areas of lawn that Pam felt were neither realistic nor sustainable in the long, dry Texas summers. Initially she kept one small patch of grass where the kids could throw a ball, before conceding that it was easier to take them to a park. One section at a time, Pam began to remove the lawn and reimagine each space. In some instances, she replaced the thirsty St. Augustine grass with shade-tolerant Berkeley sedge (*Carex divulsa*), preferring its slightly longer, shaggy look to the regular habit of native Texas sedge (*C. retroflexa* var. *texensis*), not to mention its no-mow, low-water, easy-going attitude. Other areas of lawn were replaced with sun-loving groundcovers, and all the planting beds were enlarged to accommodate an array of native and adapted plants selected for their architectural interest and ability to frame pleasing views. The last lawn holdout was a semicircular patch under a mature live oak. When the tree succumbed to canker recently, Pam knew the exposed lawn would burn in direct sun and so decided to replace it with evergreen Scott's Turf sedge (*C. retroflexa* 'Scott's Turf'), which both hides the oak's tenacious suckers and offers a softer look than the original turf. This no-mow meadow is being monitored closely for signs of heat stress.

Ongoing maintenance is mostly a matter of keeping plants within bounds, especially during the prolific growth periods of spring and early fall, but Pam loves to prune, trim, and shape her plants so she doesn't see this as a hardship. The only other significant gardening task is hand watering new plants through the summer, since city-imposed watering

TOP A dry streambed of large rounded pebbles framed with limestone boulders diverts water away from the home during heavy rainstorms. This gently winding watercourse doubles as an informal path during the dry season.

ABOVE Pam replaced the shaded lawn in this newly terraced area with easy-care Berkeley sedge (*Carex divulsa*), transitioning in sunnier areas to gray-woolly twintip, which drapes over and softens the edges of the rock wall.

Generous limestone stepping stones carve a path through a swath of deer-resistant variegated flax lily (*Dianella tasmanica* 'Variegata'). The lattice-topped gate and fence prevent deer from entering the back garden.

restrictions limit the use of a sprinkler system to once a week. One day, Pam hopes to install a rainwater catchment system to mitigate this challenge.

Yet both David and Pam acknowledge this is not going to be their forever home. Once the kids are grown, David envisions living in a condo downtown, and Pam would be happier with a much smaller garden. "I'm glad we've had this experience, though, and I'm not ready to leave quite yet. I'm just enjoying every season, knowing that we won't always be here."

Front Garden

While the interior of the home features rich jewel tones, Pam tried to restrain her penchant for bold colors in the front garden, selecting neutral gray or black pots and a more soothing, monochromatic color scheme, in keeping with the home's exterior. In a hot, dry climate, silver and gray foliage tend to dominate, so Pam chose many native and adapted plants for her front garden in this color family. Variety was added with golden variegated yuccas, deep burgundy grasses, and brightly colored flowers from shrubs and perennials, including golden thryallis (*Galphimia gracilis*) and Mexican bush sage (*Salvia leucantha*).

Using a creative blend of native and adapted plants together with a few choice annuals, Pam has developed a contemporary-traditional style that couldn't possibly be described as naturalistic: "I'm very much about *not* letting the plants do their own thing!" With a distinct focus on foliage textures, this garden is a departure from her previous flower-filled designs. Although this is due in part to the shadier conditions, Pam was also excited to try something different and incorporated many of her favorite architectural plants, including agaves and yuccas. Foliage also enables her to ensure the garden is rich with interest throughout the year, since floral displays are relatively fleeting. It goes without saying that all selected plants must withstand searing high temperatures, flash floods, extreme drought, caliche soil, and hungry deer.

Various ornamental grasses are incorporated throughout the front garden, from finely textured muhly grasses to the bold, broad blades of fountain grass (*Pennisetum purpureum* 'Vertigo'), as these have all proven to be reliably deer-resistant. Pam sites taller varieties where height is needed, and those with airy seed heads where they glow in the late evening sun.

One of Pam's favorite evergreen grasses for the shade is Sparkler sedge (*Carex phyllocephala* 'Sparkler'), whose crisply variegated green and white foliage ensures it is visible even in limited sunlight.

TOP A planter featuring bromeliads (*Dyckia* 'Burgundy Ice') receives luminescent backup from the feathery foliage of bamboo muhly (*Muhlenbergia dumosa*), and in the late light, the die-cut detail of the red metal heart, a favorite accent piece, is revealed.

ABOVE Turk's cap (*Malvaviscus arboreus* var. *drummondii*) and pale pavonia (*Pavonia hastata*) create an informal tumble of blooms among the dwarf palmetto (*Sabal minor*), foxtail fern (*Asparagus densiflorus* 'Myersii'), and Sparkler sedge (*Carex phyllocephala* 'Sparkler').

RIGHT The bold paddles of a spineless prickly pear (*Opuntia ellisiana*) echo the color of gopher plant (*Euphorbia rigida*) in the foreground, while yuccas and Mexican feather grass (*Nassella tenuissima*) weave between the two. An upright rosemary (*Rosmarinus officinalis*) and silver Mediterranean fan palm (*Chamaerops humilis* var. *argentea*) create a monochromatic yet texturally rich backdrop.

Where many designers might contrast this with bolder leaf shapes, Pam has successfully mixed it with plants that share a similar texture yet offer a distinct color.

Unfortunately, deer don't limit themselves to merely eating plants, and Pam was not prepared for the "antlering" damage, as it is known locally. "It was so frustrating," she admits. "Plants would just get to a good size and be reliably drought tolerant, only to be damaged by bucks rubbing their antlers on them." She now uses discreet cages fashioned from metal fencing panels during the fall and winter rutting season to protect vulnerable agaves, hesperaloes, yuccas, and small trees.

Artistic Expression

Throughout the property, Pam has used art to add color, establish a focal point, or introduce unexpected detail. Two distinct themes within her art collection are also evident: hearts and spheres. Pam admits these themes evolved over time. There was no master plan; she simply purchased pieces that appealed to her. On the eastern side of the house a path of decomposed granite leads past a secluded

TOP LEFT Sections of bent wire fencing are used to protect a giant hesperaloe (*Hesperaloe funifera*) from antler damage.

TOP RIGHT Gregg's blue mistflower (*Conoclinium greggii*) attracts many butterfly species, including the Monarch.

ABOVE Inland sea oats (*Chasmanthium latifolium*) and bamboo muhly, incandescent in the sunlight, frame a deer-resistant gate adorned with three hanging hearts that twirl in the breeze.

LEFT Dwarf palmettos flank a tall container featuring a thriller (×*Mangave* 'Pineapple Express') and a spiller (*Dichondra argentea* 'Silver Falls'). A Japanese plum yew (*Cephalotaxus harringtonia* 'Prostrata') and golden sedge add evergreen foliage at the base.

side garden to the heart gate. In the side garden's sitting area, a rustic bench decorated with metal bird silhouettes provides the perfect place to watch butterflies as they visit the sprawling mounds of Gregg's blue mistflower (*Conoclinium greggii*). Behind this, the brick garage wall has been updated with dark-stained trellises installed over mirrors, ingeniously suggesting windows. This side garden also features a number of spheres, their perfect form signifying completeness, a popular symbol in landscape design.

The most conspicuous spheres, due to their elevated position in the front garden, are the three blue ceramic balls nestled into the carpet of gray-woolly twintip, their color echoing that of a nearby tuteur. Containers too are used as an art form throughout the front garden. Where some homeowners might be satisfied with a token pot or two by the front door, Pam has incorporated them both to beautify and to problem-solve. For example, to break up the home's brick façade, Pam introduced a sleek gray pot. Realizing that it would take time for its occupant, a young mangave (×*Mangave* 'Pineapple Express'), to establish, she added a fun metal sculpture to create an immediate sense of stature. In another shaded pathway adjacent to the garage, where no in-ground planting was possible, stone pillars were used to elevate short containers, adding interest, texture, and color.

In a challenging region, where so many watch helplessly as their traditional lawns and foundation plantings shrivel under the relentless sun, Pam's garden is an achievement—and she is justifiably proud of it. Interestingly, a few neighbors seem to be taking a fresh look at their own gardens, and new families are moving in and reviving the neighborhood. Pam's influence and inspiration must surely be an encouragement to any homeowner aspiring to create their own beautiful oasis.

Top 10 Plants

BAMBOO MUHLY (*Muhlenbergia dumosa*). With stems that are reminiscent of bamboo and finely textured, feathery foliage that sways gently in the slightest breeze, this ornamental grass is deservedly popular. Drought tolerant once established and usually evergreen, bamboo muhly makes an attractive living screen or container specimen, 3–5 feet tall, 3–4 feet wide. Full sun–partial shade, zones 7–10. Deer resistance: A.

VERTIGO FOUNTAIN GRASS (*Pennisetum purpureum* 'Vertigo'). A bold, architectural grass, 4–6 feet tall and half as wide, whose broad green blades turn deep purple as the summer progresses. An annual in most areas, this ornamental grass can be used in the landscape or containers. Full sun, zones 8–11. Deer resistance: A.

TEXAS SOTOL (*Dasylirion texanum*). An exceptional evergreen succulent for hot climates and full sun; an architectural sculpture in the landscape or containers. The narrow green leaves have spiny margins, a good deterrent against all manner of critters. Plants are 5 feet tall and wide; a 9- to 15-foot-tall spike bearing cream flowers may appear in summer. Drought tolerant. Zones 7–10. Deer resistance: A.

PROSTRATE JAPANESE PLUM YEW (*Cephalotaxus harringtonia* 'Prostrata'). Tolerant of shade, heat, drought, and deer, this evergreen conifer is both useful and versatile. The attractive, dark green needled foliage can be used as a low hedge or a groundcover, 2–3 feet tall, 3–4 feet wide. Partial shade–full sun, zones 6–9. Deer resistance: A.

SILVER MEDITERRANEAN FAN PALM (*Chamaerops humilis* var. *argentea*). With icy blue, fan-shaped foliage with silver undersides, this multi-trunked palm makes a truly eye-catching specimen in the landscape. The overall form is round as it matures, to 8–12 feet tall and wide. Full sun, zones 8–10. Deer resistance: A.

GOLDEN THRYALLIS (*Galphimia gracilis*). A fast-growing shrub, 4–6 feet tall, 4 feet wide, and evergreen in milder climates. Bold yellow flowers adorn the plant from spring until frost. Drought tolerant once established. Full sun–partial shade, zones 9–11. Deer resistance: A.

TURK'S CAP (*Malvaviscus arboreus* var. *drummondii*). This deciduous native shrub has an informal habit and a profusion of bright red, turban-like flowers that attract bees, butterflies, and hummingbirds. Grows 3–9 feet tall, 3–5 feet wide. Thrives in full shade but adaptable to full sun. Zones 7–11. Deer resistance: A.

PALE PAVONIA (*Pavonia hastata*). A Texas native, 5 feet tall, 7 feet wide. Light pink, hibiscus-like blooms, each with a deep burgundy eye, adorn this woody perennial from spring until fall. Evergreen in warmer climates, hardy but deciduous in cooler areas. Full sun–partial shade, zones 8–11. Deer resistance: A.

PALE-LEAF YUCCA (*Yucca pallida*). Like all yuccas, this Texas native requires good drainage. The broad blue-green succulent leaves have a pale yellow margin, and if grown in full sun, a tall flower spike of white bell-shaped blooms may be produced in late spring. Deer may be tempted to eat the blooms occasionally but do not eat the foliage. Grows 1–2 feet tall and wide. Full sun–partial shade, zones 8–10. Deer resistance: A.

GREGG'S BLUE MISTFLOWER (*Conoclinium greggii*). A native, herbaceous perennial, 1.5 feet tall, 2 feet wide, this spreading groundcover is a good filler in the garden. Butterflies love the fluffy lavender flowers that cover the plant in summer and fall. Full sun–partial shade, zones 7–10. Deer resistance: B.

A Blue Jeans Garden
↳ where kids and dogs are always welcome

Anna and Todd Brooks envisioned a casual home and garden where everyone would feel welcome—with the exception of the deer, who would be encouraged to pass on through. Over twenty years later their dream has been realized, thanks to hard work, intuition, experimentation, and good design.

Both Anna and Todd hail from families of enthusiastic gardeners. Anna's great-grandmother was an herbalist who immigrated to the United States from England; Todd's mom, Dot Hollerbach, founded the award-winning landscape design company Arcadia Gardens LLC, based in Michigan's Great Southwest, which Anna and Todd now own and run. In fact, Dot takes full credit for introducing Todd to his future wife when she hired Anna, a graduate from the landscape design program at Michigan State University. A knowledgeable horticulturalist, Anna

QUICK FACTS
LOCATION: Stevensville, Michigan (zone 6)
SOIL TYPE: sandy loam
PROPERTY SIZE: 1.3 acres
PROBLEM CRITTERS: deer, rabbits, voles, squirrels, groundhogs, turkeys
OTHER CHALLENGES: extremely dry, sunny, open lot; no irrigation

DESIGN CREDITS:
home of Todd and Anna Brooks
designed by Todd and Anna Brooks, Arcadia Gardens LLC

Tucked away at the end of a quiet cul-de-sac, set amid towering native trees and informal gardens, is the charming house that the Brooks family calls home.

brought an artistic eye for color and detail to the team, and a caring, thoughtful nature that made her very popular with clients. Like Todd, she also had a blue heeler for a canine companion at the time. With so many connections and shared interests it was no surprise when the two married in 2000.

Todd is a self-described "goals guy," and when he went looking for a home in 1996 for himself and his young daughter, Emily, he had a very short wish list: a child- and dog-friendly space, a walkout basement, and a blank-slate landscape with a slight modulation in grade, in order to realize his dream of creating a sunken garden. The home he found came with an interesting history, having been part of the House of David, a Michigan-based religious colony that owned around a thousand acres in the early twentieth century (one of the original bunk rooms remains on the property, now functioning as a storage shed). By the time Anna moved in four years later, Todd had established the bones of their future landscape, putting his skills to good use as he crafted the initial paths, retaining walls, and patios out of stone. In the years since, a front porch and two side extensions have been added to the original 900-square-foot house to accommodate the growing family, which now includes teenage sons Camden and Nathan.

Samson, a friendly blue heeler, completes the Brooks family today. When he's not snoozing on the shady front porch, he's chasing or hunting for balls, his incursions frequently sending him through billowing geraniums (Geranium sanguineum 'Max Frei'), which flatten briefly under his sturdy frame but quickly recover.

Over time, Todd and Anna have transformed this wooded lot into a delightful, sunny garden where one immediately feels at ease, with space for everyone within an informal framework of colorful flowers, many of which have been passed along by clients and friends. Not all their ideas have been successful, however. Todd laughs as he recalls planting a small orchard ("We are not fruit farmers"); Anna adds that the necessary pruning and spraying were just not worth the investment of time for the meager colander of worm-ridden cherries they harvested every fifth year, or the one paltry peach that turned out to be bad. And their initial vegetable garden design, although beautiful, was too large, an over-calculation of the family's needs and available time to manage it. Another problem was that invasive Bermuda grass from the driveway was constantly overrunning the plot, so Anna and Todd brought in an excavator to remove soil to a depth of 2 feet, replaced it with fresh sandy loam, and then reconfigured the space to better suit their needs.

While Anna admits she loves perennials and grows as many of her favorites as possible, this garden also serves as an important testing site for plants she is considering using for clients. Her rigorous assessment for deer resistance, minimal watering (there is no irrigation in her garden and hoses are used sparingly), fast-draining sandy soil, and variable Michigan winters quickly eliminates the weaker, less reliable specimens. Anna also uses her garden to experiment with new combination ideas gleaned at design conferences. Yet function must come before beauty. "Plants have to be cast iron," says Anna. "We have a lot of kids and dogs visiting." And deer.

Planning the Garden

A herd of five to ten deer call this their home, although only one or two are usually seen at a time. Having lived here for so long, the Brookses are familiar with the paths the deer take through their property and designed their landscape accordingly. Their primary tactic is to keep known deer trails free of ornamental plants in order to funnel deer through the space, encouraging them to keep moving rather than offering an invitation to stay and browse. The most challenging area in this regard was the section of driveway that the deer crossed daily in order to reach the path that leads them down the hillside. While the front garden is designed for the most part with deer-resistant plants, it is also the primary area for growing vegetables. Dense, layered plantings mostly hide the edibles from passing deer, although in spring there may be some grazing on the plants closest to their path.

Anna seldom uses deer repellent sprays in her garden, preferring to rely on plants selected for their known deer resistance; however, when a special

The country home is set back from the wide, circular gravel driveway; a lavender-lined pathway leads to the back garden.

event is planned, she might use Liquid Fence on susceptible plants, such as her Knock Out roses. While a tall deer fence would afford the ultimate protection, it would compromise the borrowed views the family enjoys of neighboring gardens as well as restricting the passage of smaller wildlife, such as coyotes and badgers.

Deer notwithstanding, the garden is designed to offer something of interest in each season and exhibits several different waves of color during the year. Spring sees a glorious explosion of purple as rhododendrons, lilacs, and eastern redbuds (*Cercis canadensis*) bloom, complemented by hundreds of daffodils and several dwarf forsythias (*Forsythia* 'Gold Tide'). Mature spirea hedges surround the property and line the deer paths, their sprays of white blooms attracting bees and butterflies but mercifully not the deer.

Several varieties of peonies and irises herald the start of warmer weather, while daylilies, tickseeds, false indigo (*Baptisia australis*), and Russian sage (*Perovskia atriplicifolia*) are just some of the colorful perennials to celebrate summer.

Many trees and foundation shrubs anchor the fiery autumnal display, with aromatic aster (*Symphyotrichum oblongifolium*) and goldenrod (*Solidago rugosa* 'Fireworks') being the final perennials to bloom before winter. Under a snowy blanket, evergreen conifers are the main foliage contributors; native witch hazels (*Hamamelis virginiana*) add spidery, fragrant yellow blooms in late November and early December.

Today the Brooks garden is a series of interconnected rooms, each with its own role. At the front of the home, the wide gravel driveway loops around a teardrop-shaped planting bed, its broadest part laid out as the circular vegetable garden. The center of the vegetable garden is planted with mint, thyme, sage, and other herbs; the outer borders include tomatoes, peppers, and squash, interplanted with

OPPOSITE The vegetable garden makes good use of the island in the front driveway but is screened from passing deer by mature trees and shrubs.

TOP LEFT Fritsch spirea (*Spiraea fritschiana*) has proven to be an easy-care shrub for flanking known deer paths.

TOP RIGHT Assorted daylilies (*Hemerocallis*) and black-eyed Susans (*Rudbeckia*) steal the show in summer, sizzling against the cooler gray-blue of the house and the blue and silver spikes of Russian sage (*Perovskia atriplicifolia*).

LEFT The large, felted leaves of lamb's ears (*Stachys byzantina*) weave through the lower twiggy branches of a slender deutzia (*Deutzia gracilis* 'Nikko'), highlighting the shrub's white spring blooms and extending the season of interest.

zinnias and marigolds. Buffering this tasty treat from the view of inquisitive deer is a mature Canadian hemlock (*Tsuga canadensis*), a purple-leaf sand cherry (*Prunus* ×*cistena*), and several white-blooming slender deutzias (*Deutzia gracilis* 'Nikko'), all edged with silvery lamb's ears (*Stachys byzantina*). As with the rest of the garden, seedlings are encouraged, so a few columbines (*Aquilegia vulgaris*), received as a gift from a client years ago, have multiplied into a delightful, ever-expanding carpet, mingling with drifts of daffodils and grape hyacinths (*Muscari armeniacum*) for a colorful spring display. Edible rhubarb has proven reliably deer-resistant, and its large leaves add a wonderful focal point to the island bed; strawberries edge the outer perimeter.

Flanking the front porch are two beds that give a nod to Todd's preferred design style—symmetry and order—with redbud trees anchoring each space, underplanted with large swaths of big blue lilyturf (*Liriope muscari*), bearded iris, and lavender. These borders are easy to reach with a hose when needed, so include a few of the thirstier perennials, such as astilbes and leopard plant (*Ligularia*). A compact, bushy lavender (*Lavandula angustifolia* 'Munstead') reseeds in the sandy soil with abandon and can withstand rogue basketballs from the nearby hoop, making it an ideal candidate for edging this border.

A series of wooden and brick steps leads off from the south side of the house down to its lower level, which used to serve as a home office but has recently been recommissioned as a man (and boy) cave as well as occasional guest quarters. A small patio outside the sliding doors is made from reclaimed bricks salvaged from a local road construction project. Plantings in this area are kept simple and reliably deer-resistant, since this is the most highly trafficked area. Tall native trees are underplanted with Fritsch spirea, witch hazel, western red cedar (*Thuja plicata* 'Zebrina'), and American cranberrybush (*Viburnum*

LEFT Borrowed views of distant trees add to the sense of spaciousness on the Brooks property.

BELOW LEFT The treehouse was designed by Todd to mimic the colors and architecture of the home.

BELOW The faded white flowers of an oriental hellebore (*Helleborus orientalis*) play off the silvery Japanese painted fern (*Athyrium niponicum* var. *pictum*). Both these deer-resistant perennials thrive in Michigan's climate.

OPPOSITE In early summer, the sledding hill is transformed into a painterly meadow as dame's rocket (*Hesperis matronalis*) is encouraged to self-seed among the grasses.

trilobum 'Wentworth'). At the rear of the home, a delightful shade garden includes Japanese painted fern (*Athyrium niponicum* var. *pictum*) and oriental hellebore (*Helleborus orientalis*), the species doing better in Michigan's harsh climate than the newer hybrids in Anna's experience. A treehouse, designed and built by Todd, perches above this garden, long since abandoned by the boys but offering a fun hangout spot for squirrels and chipmunks, which are often seen scurrying about on the platform.

Beyond this shady border, the land drops off sharply, a feature that the boys put to good use in winter months for exhilarating sled rides. In late May and early June, however, this sledding hill is an impressionistic matrix of lavender and white dame's rocket (*Hesperis matronalis*) planted amid native grasses. This area gets mowed just once a year, after

the dame's rocket has set seed. By fall the hillside is blooming again, this time with purple fall asters. The sheer abundance of these blooms together with their notable deer resistance maintains an acceptable balance between the informal meadow-style planting and a habitat where the deer are welcomed, with minor damage barely noticed.

    The family takes most of their meals in the large, screened porch that overlooks this picturesque hillside. It is a magical place in the evening, when the soft glow of string lights illuminates the space. Once the trees lose their leaves, the Brookses can more easily observe the herd of deer as they amble down the slope or use the grasses for bedding down in spring with their young. From here, a simple path connects to a short flight of steps that leads into the back garden.

Back Garden

This well-proportioned area features a sunken firepit, two patios, and a large lawn, where a soccer ball can be kicked without concern and Samson is free to enjoy hours of fetching sticks and balls. From the garage at the north end of the driveway, a flagstone path winds between a native, self-sown tulip tree (*Liriodendron tulipifera*) and a Canadian hemlock, leading to the large lawn. A buffer of lilac and bridalwreath spirea (*Spiraea ×vanhouttei*) separates this upper garden from a sunken patio, which is reached by a flight of curved stone steps, each one a little wider than the last during descent. This lower garden is encircled by a low, dry-stacked stone wall that curves to create a circular sunken firepit patio with built-in seating, a favorite spot for casual entertaining; and a small patio attached to the firepit is

ABOVE Bridalwreath spirea (*Spiraea* ×*vanhouttei*) grows into a large, arching shrub that the Brookses use to divide garden spaces.

LEFT The flight of curved steps is flanked by one of Anna's favorite ornamental grasses, *Miscanthus sinensis* 'Morning Light'.

furnished with a weathered picnic table for al fresco dining at any time of day. A wide corridor of grass links the sunken garden to the upper lawn and the small storage shed that used to be a bunk room.

There are far more birdfeeders than garden art in the Brooks garden, which in hindsight is just as well. A few years ago, Anna treated herself to a beautiful glass gazing ball in shades of purple and blue, and set it carefully in the garden. Coming home after a long day at work, Todd assumed it was one of the kids' balls that had been abandoned, and promptly kicked it out of the border, whereupon it smashed to smithereens. Poor Todd had some explaining to do when he ventured into the house.

Managing the garden is now a family affair, with Todd continuing to design and build the hardscape features, and Anna in charge of selecting the right plant for the right place—a key tenet of reducing maintenance in a garden. This means taking into account the plant's mature size and the cultural conditions it needs to thrive as well as the anticipated level of deer resistance. Meanwhile, Camden has been assigned the job of mowing the lawn, and Nathan, the youngest member of the family, loves

to grow anything he can eat, his current favorite being fat, juicy strawberries. It's all a shared labor of love—yet one that does not seem so laborious. To sum up the Brooks philosophy: accept each moment with gratitude, receiving it as a gift to be cherished. Rather than worrying about stray plants, playful dogs, energetic kids, or inquisitive deer, they have designed a welcoming, casual garden where all can thrive in harmony, each in their own space.

LEFT The gazing ball that now has pride of place next to the blue false indigo (*Baptisia australis*) is a resilient aluminum one, received as a gift and penance from Todd one Mother's Day.

BELOW The back garden is divided into two distinct functional areas: while kids and dogs play on the upper level, parents can enjoy time by a bonfire.

Top 10 Plants

LAMB'S EARS (*Stachys byzantina*). Thriving in hot, dry conditions and full sun, this low-growing perennial, 1.5 feet tall and wide, is beloved for its velvety soft silver foliage. Short, fat spikes of pink flowers appear in summer; many gardeners prefer to remove these to maintain a tidier appearance. Zones 4–8. Deer resistance: A.

DESDEMONA LEOPARD PLANT (*Ligularia dentata* 'Desdemona'). An imposing, clump-forming perennial, 2–3 feet tall and up to 2 feet wide, which is grown as much for its foliage as for the large yellow daisies, which bloom in summer. Needs amended, moisture-retentive soil in partial shade–full shade to thrive. Lack of water and direct sunshine will both result in wilting. Bait for slugs. Zones 3–8. Deer resistance: A.

MAX FREI GERANIUM (*Geranium sanguineum* 'Max Frei'). Abundant magenta flowers cover this bushy perennial in May and June, with intermittent reblooming until frost. This variety has shown better heat and cold tolerance than many hardy geraniums, and although Rutgers notes it is occasionally severely damaged by deer, that may serve simply to reduce self-seeding and keep the perennial trimmed to shape. Grows 1 foot tall, 2 feet wide. Full sun–partial shade, zones 3–8. Deer resistance: C.

LORELEY IRIS (*Iris* 'Loreley'). This 26-inch-tall heirloom bearded iris is wonderfully showy, with primrose-to-amber standards and richly veined violet falls. Full sun, zones 3–8. Deer resistance: A.

WALKER'S LOW CATMINT (*Nepeta racemosa* 'Walker's Low').
Loose stems of lavender-blue blooms are held high above the clumping, aromatic gray-green foliage. Shearing the spent flower spikes encourages rebloom, keeping bees and floral arrangers happy. This variety is sterile (so self-seeding is not a problem), but plants are easily propagated by division. Grows 2.5 feet tall, 3 feet wide. Full sun–partial sun, zones 4–8. Deer resistance: A.

FALSE INDIGO (*Baptisia australis*).
In May and June, lupin-like blue flowers stand well above the blue-green foliage of this perennial. Native to the eastern United States, it is noted for its drought tolerance and ability to naturalize in meadows and prairies as well as in more informal garden settings. Grows 3–4 feet tall and wide. Full sun–partial shade, zones 3–9. Deer resistance: A.

COLUMBINE (*Aquilegia vulgaris*).
These spring-blooming perennials, 1.5–3 feet tall and up to 2 feet wide, tolerate full sun but do best in partial shade, where they may self-seed prolifically—sometimes to the point of being a nuisance. Hummingbirds are attracted to the true-blue blooms. Zones 3–8. Deer resistance: B.

MUNSTEAD LAVENDER (*Lavandula angustifolia* 'Munstead'). A compact, early-blooming variety of English lavender, 1.5 feet tall and wide. This semi-woody perennial is known for its lavender-blue flowers, which are popular for culinary purposes. Full sun, zones 5–8. Deer resistance: A.

ZAGREB TICKSEED (*Coreopsis verticillata* 'Zagreb'). The mounding cushions of thread-like, green foliage, 2 feet tall and wide, are attractive in their own right, but when studded with dozens of rich golden yellow daisies in midsummer this perennial truly lights up the garden. An extremely drought-tolerant addition to the deer-resistant border. Full sun, zones 3–9. Deer resistance: A.

CAESAR'S BROTHER IRIS (*Iris* 'Caesar's Brother'). Siberian irises will tolerate a wide variety of soils but need to be divided regularly for best bloom production. Caesar's Brother is a popular variety, 3–4 feet tall, 2.5 feet wide, known for its velvety, deep purple flowers, which are excellent in floral arrangements. Full sun–partial shade, zones 3–8. Deer resistance: A.

A Storyteller's Garden
↳ unveiling magic, one chapter at a time

With a cantankerous green-winged macaw, a Moluccan cockatoo that "spoke in tongues," a sweet (unless hormonal) eclectus parrot, a basset hound called Georgia Mae, and a grand piano to find a home for—Jay Sifford wasn't the typical house hunter. But as his realtor, Jay's mom understood his needs better than most and decided to take a gamble when she showed him a rundown house on a steeply sloping wooded lot. Jay immediately felt a connection to the tree-top home, which reminded him of a mountain cabin, and in 1998 he moved in. After a few years focused on remodeling the interior, Jay finally turned his attention to the exterior, with neither a plan nor a vision. Originally there weren't even steps up the steep slope to the front door, and the landscape was an obstacle course of dilapidated railroad ties, sticky red mud, and endless liriope. And then there were the deer—up to thirteen every day.

QUICK FACTS
LOCATION: Charlotte, North Carolina (zone 8)
SOIL TYPE: loam and clay
PROPERTY SIZE: 0.5 acre
PROBLEM CRITTERS: deer, squirrels, chipmunks, skunks, raccoons
OTHER CHALLENGES: steep hillside; erosion; no irrigation; extensive shade

DESIGN CREDITS:
home of Jay Sifford
designed by Jay Sifford; Sifford Garden Design

Until moving to this garden Jay knew little about deer. Not only was their voracious appetite for his favorite hostas, coral bells, and hydrangeas an unwelcome surprise, but so was their habit of rutting, with young trees snapped in half and many others left disfigured. The learning curve was almost as steep as the land, and Jay discovered by painful trial and error which plants were reasonably deer-resistant. Although a deer fence would solve the problem, it wasn't a practical solution since that would have meant installing an 8-foot-tall gate across the steep driveway (local codes set a limit of 6 feet), and the hilly terrain posed a significant challenge. And deer weren't the only visitors. Squirrels and chipmunks frequently created mischief, as did the skunks, which dug up all Jay's newly planted shrubs—every day for thirty days—before eventually growing tired of their game.

Dealing with wildlife was not the only challenge. Gardening on a slope meant learning about erosion control and water runoff. Failures were common as Jay learned to interpret the subtle nuances of shade cast by the canopy of beech (*Fagus grandifolia*) and sweetgum (*Liquidambar styraciflua*) and tested plants in an effort to follow basic plant tag information, which seldom went beyond stating a plant's preference for sun or shade. North Carolina's high humidity further complicated the selection of appropriate plants; rarely do nursery tags tell the gardener whether a plant can take high humidity, or if it will simply melt away in a diseased puddle by mid-August.

To gain a better understanding of the elements that go into designing a cohesive landscape, Jay began watching gardening programs on television, gaining "just enough knowledge to be dangerous." But his real inspiration came from books. *The Inward Garden* (Bunker Hill Publishing, 1995) by Julie Moir Messervy was perhaps the most influential, encouraging readers to consider both the natural world and the contemplative spirit within, a philosophy that resonated deeply with Jay. Another firm favorite is *From Art to Landscape* (Timber Press, 2010) by W. Gary Smith, which discusses how to combine the essential character of a place with one's innermost creative spirit. As a psychology major with an interest in philosophy and contemporary art, Jay never imagined that he would one day put all those

OPPOSITE Setting the tone: besides matching the colors of the surrounding foliage, the various forms of a totem pole by local artist Kimberly Tyrrell pique one's curiosity and invite interpretation.

LEFT The front entry garden is brightened by a chartreuse smokebush (*Cotinus coggygria* 'Golden Spirit'), a soft peach and green variegated sycamore maple (*Acer pseudoplatanus* 'Eskimo Sunset'), and the last few blooms of a peony (*Paeonia* 'Kopper Kettle').

ABOVE The gravel path serves as a negative space, somewhere for the eye to rest amid the tapestry of color and texture.

A Storyteller's Garden

qualities to use as a landscape designer. In fact, he hesitates to call himself a designer at all, preferring to be known as an artist, magician, storyteller, and horticulturalist—in that order. Rather than seeking credit for designing this intriguing garden, Jay feels his role has been to reveal the stories it wanted him to tell, and he recommends exploring the garden with a childlike sense of wonder and imagination. Over time he has created several garden rooms within this botanical wonderland, likening each to one of the sidewalk chalk drawings in *Mary Poppins*. Every space becomes a magical new reality for the adventurous travelers as they step inside. All are designed to be experienced and savored at a leisurely pace, fully engaging the senses while encouraging discovery and personal interpretation.

The half-acre lot drops away dramatically into a creek, leaving the home perched like a treehouse high above the forest floor with every window and deck offering a vista of thought-provoking art installations amid abundant plantings. As Jay's appreciation for blue-needled conifers grew, he decided to repaint the home a deep steel-gray, accented with a light blue chimney and burgundy trim, effectively connecting the home to the landscape. The deer-resistant plant palette has been selected for year-round interest, anchored by assorted conifers and other evergreens (edgeworthia, sarcococca, yellow anise); deciduous trees and shrubs are included to celebrate the change of the seasons. Daylilies are one of the few flowers; Jay enjoys their fleeting blooms and accepts occasional taste-testing by the resident herd.

The original struggling, shady front lawn has been replaced by a wide gravel path flanked by colorful foliage plants in a new front entry garden. This path is up to 8 feet wide in places, to accommodate plantings that spill beyond their allotted boundaries. Shallow steps along its course force a slower pace, a deliberate strategy to engage visitors in their surroundings, reinforced by the pea gravel that shifts underfoot. The front entry garden displays the most vibrant plant selection with chartreuse, burgundy, and variegated foliage mingling with shades of blue and green. As one moves into the more secluded, shadier spaces, brighter colors are edited out, until finally, in the outermost meditation circle, color is reduced to its purest essence—that of light.

The path then sweeps under the feathery green archway of a weeping bald cypress (*Taxodium distichum* 'Cascade Falls') with steps down into a conifer amphitheater, a natural depression that showcases Jay's ever-expanding collection of choice conifers that thrive in the fertile soil, enriched over many years with leaf mold. Beyond an ancient boulder one steps onto a Japanese yatsuhashi boardwalk, painted a traditional Chinese red and a key focal point in the landscape. From this vantage point one can look out onto the fern glen and creek and catch glimpses of the light garden in the distance.

Closer to the house, at the end of the driveway, a tiered waterfall and koi pond have been carved into the hillside. This was the first garden project to be undertaken, piquing the interest of neighbors, fellow gardeners, and eventually garden tours. Steep bluestone steps to one side lead to the home's entrance. The modest width of the path encourages a slower pace, as visitors find themselves brushing up against the glossy leaves of Japanese aralia (*Fatsia japonica*), or treading with care to avoid an errant stem of strawberry saxifrage (*Saxifraga stolonifera*) that creeps between the steps. Continuing a little farther uphill, above the uppermost waterfall, one passes through antique Chinese doors to a small, shaded outdoor room where a red Chippendale bench invites one to pause; this upper woodland garden is the highest point on the property. The most recent garden to be created is the light garden, a meandering pathway through the forest that celebrates sunlight and shadows, and serves as an outdoor gallery for several metal sculptures, culminating in a circular meditation garden.

Jay insists his garden is fairly low-maintenance, the primary chore being managing leaf litter in

A weeping bald cypress (*Taxodium distichum* 'Cascade Falls') marks the transition from the upper front entry garden into the conifer amphitheater. An installation of red bamboo by glass artist Jesse Kelly accentuates the portal.

fall, his aversion to noisy leaf blowers necessitating the use of a rake. Lack of irrigation also means that he spends time watering, "with a glass of wine in one hand and a hose in the other." How long does he water? "By the time I've had three glasses, we're all good!" To Jay's surprise, neither the deer nor the terrain proved to be the greatest challenge in the end; the real hill to climb was figuring out how he related to the land, and in the process, he became a completely different person. Jay's discerning eye for detail and horticultural artistry have resulted in him becoming a designer much sought-after by those wishing to experience transformative magic in their own gardens. Clearly he has transcribed the untold stories of this land into an engaging book that captivates the imagination of all who care to read it.

Front Entry Garden

After removing the sparse, sun-deprived lawn, Jay planted an extensive collection of specimen Japanese maples and conifers on either side of a wide gravel path, transforming the front garden from predictable to remarkable. Foliage in shades of blue and burgundy predominate, accented by several golden conifers, including a shore juniper (*Juniperus conferta* 'Golden Pacific') and an oriental spruce (*Picea orientalis* 'Skylands'). Many of the conifers display distinctive weeping or prostrate forms, and Jay readily admits there are far more specimen conifers in this small space than he would ever introduce in a client's garden, yet it works. "I like trees that have elevated arthritis to an art form," he says of their gnarled, contorted shapes. "They unlock people's imagination." Indeed, the undulating, prostrate Big Wave Norway spruce (*P. abies* 'Big Wave') suggests a sea creature, or perhaps a giant green spider, stealthily maneuvering its way across the garden in search

A Norway spruce (*Picea abies* 'Big Wave') is combined with coral bells (*Heuchera* 'Venus') and a mounding Japanese white pine (*Pinus parviflora* 'Tanima-no-yuki'); a Rocky Mountain juniper (*Juniperus scopulorum* 'Tolleson's Blue Weeping') drapes overhead, creating a striking monochromatic medley.

The branches of the weeping bald cypress (*Taxodium distichum* 'Cascade Falls') are selectively trimmed to create a scrim effect.

of unsuspecting prey. This garden challenges the mind to rethink preconceived ideas of what a tree might look like, what Jay describes as a "metaphorical cleansing of the mind" before entering into the other areas of the garden.

Several large rustic containers punctuate the plantings, their perceived age adding to the sense of history, while the colors of the ceramic glazes echo the surrounding foliage. To enhance the experience and encourage interaction, one frequently has to brush rogue branches or tendrils aside in order to pass. Stepping through the archway created by a weeping bald cypress one has to part the trailing strands of feathery leaves that hang like a beaded curtain, creating a sense of anticipation upon entering the conifer amphitheater, which serves as a transitional space in the journey to the Asian-inspired boardwalk. Although this garden has color year-round, Jay enjoys it most in spring when the Japanese maples (*Acer palmatum*) first leaf out; from deep burgundy ('Rhode Island Red') through delicate greens (Dissectum Viride Group) to vivid pink variegation ('Hana-matoi'), there is always something new to marvel at.

Thankfully deer seem to prefer an easier route down to the fern glen rather than bushwhacking their way through the closely planted conifers and thorny barberries that line the drive; their browsing is usually limited to some daylilies, which quickly regrow, but their careless jumping did damage an espaliered fence of weeping blue Atlas cedars (*Cedrus atlantica* 'Glauca Pendula') recently. Jay is also careful to keep tender hostas and coral bells away from the preferred trails of passing deer so as not to pique their interest.

Conifer Amphitheater

There was a danger, as with all plant collections, that this area could become a disparate jumble of individual, personality-rich specimens. To avoid this, Jay created a sense of unity by repeating key colors, textures, and forms. Smaller boulders were hand-carried into the area to expand the stonework, and swaths of intensely fragrant Cheddar pinks (*Dianthus gratianopolitanus* 'Firewitch') were woven through the conifer display, their neon-pink flowers adding an unforgettable note of cloves to the spring air. Several ground-hugging Golden Pacific shore junipers have also been used to carpet the space.

Sadly, two rare, highly prized weeping cultivars of Norway spruce (*Picea abies* 'Cobra') were lost from this area as a result of deer damage. In fact, Jay has noticed that the deer seem to be especially drawn to rutting on Norway spruce and blue Atlas cedars (*Cedrus atlantica*), which is unfortunate since they are two of his favorite conifers. Yet with seventy-five Japanese maples and more than twice as many conifers, he cannot adequately barricade the plants to protect against such winter damage and has accepted there will always be some losses.

Fern Glen

The bench at the end of the Japanese yatsuhashi boardwalk is Jay's favorite spot in the entire garden. Whether savoring the morning solitude or a last glass of wine, it is here that he feels the greatest connection to the land as he watches the ever-shifting patterns of dappled light and shadow. This is the lowest point on the property, representative of a place of humility and submission, a restorative setting where Jay "feels an infusion of creative energy."

It is also an area rich in stories that Jay loves to hear and share. Diminutive army figures, long since lost by previous generations of children, tell of simpler times, while initials carved into the smooth gray bark of a beech tree make one wonder what became of their professed love. One cannot but feel humbled by the majestic trees towering overhead or be awed by the passage of time suggested by the

OPPOSITE The conifer amphitheater showcases many of Jay's favorite specimen conifers; a Hana-matoi Japanese maple in a low, curvaceous bowl adds a focal point at the edge of the path.

BELOW Painted Chinese red, the distinctive zigzag boardwalk culminates in a small deck, cantilevered over the forest floor, the centerpiece of the fern glen.

TOP A seedling of mayapple (*Podophyllum peltatum*), a North American native, pushes up through the leaf mold beside a silvery Ghost fern (*Athyrium* 'Ghost'). Both are reliably deer-resistant.

ABOVE A new chapter in the story of the fern glen begins.

lichen-encrusted boulders that pepper the ground beneath them. Many people urged Jay to carve paths through the trees and build bridges over the creek, yet he felt that would be intruding on this sacred space; so he built the zigzag boardwalk, cantilevered over the forest floor, where he might observe and listen to the sights and sounds of the forest without disturbing the flora and fauna creating them. The deep, resonant toll of oversized windchimes suspended high above the boardwalk adds to the meditative ambience.

Three and a half years ago this area was a tangled mess of poison ivy, English ivy, and honeysuckle that obscured the creek and choked the land. Jay methodically cleared the weeds and removed almost a hundred young beech saplings that were overcrowding the space, while limbing up the remaining trees as far as his pole pruners would reach to allow more light to penetrate. To date he has added some forty species of ferns, almost twelve hundred plants in total, in addition to native shrubs and various deer-resistant perennials, all of which are thriving in the dappled shade.

Designing a garden with so many ferns took considerable thought and planning. The more typical groups of three or even seven plants would be lost in such a vast, open space, which instead calls for much larger drifts. Jay also needed to pay careful attention to the small details that made each fern distinct. He assessed four key characteristics: size, shape, color, and texture. When considering whether or not to place two ferns next to one another, he looked for a pleasing balance between similarity and contrast. If all four criteria were the same, the combination would be unremarkable; when two or more aspects differed, a sense of cohesion balanced with visual interest was achieved. The focus is on texture rather than a kaleidoscope of colors. In such a shady environment, the subdued palette relies on gradations of green; the coppery new growth of autumn ferns introduces a different note in spring, and silver Japanese painted ferns (*Athyrium niponicum* var. *pictum*) add subtle contrast. Included in the glen are several evergreen ferns for winter interest, including autumn fern (*Dryopteris erythrosora*), tassel fern (*Polystichum polyblepharum*), and the native Christmas fern (*P. acrostichoides*), which Jay transplanted from other areas of his woodland. The latter has proven to be one of the most reliably drought-tolerant varieties.

Once considered an unbuildable lot with a significant creek rushing through it, the stream that runs through the fern glen is now a mere shadow of its former self, having being rerouted when neighboring homes were built. In accordance with Asian philosophy, Jay installed a series of stone pillars along the creek to act as protectors of this much-diminished life source; they also have the practical advantage of improving the creek's visibility when viewed from above. Deer visit the glen daily yet do little damage other than trampling a few fern fronds. A newborn fawn was discovered by chance, bedded down in the Christmas ferns and still moist from birth—a humbling moment for Jay as he appreciated anew that he is merely the steward of this land, not its owner.

Light Garden

What began as a heavily eroded wasteland of dead trees and invasive weeds is now a remarkable garden featuring a serpentine, dry-stacked Tennessee stone wall, several bronze and steel sculptures, and a meditation circle. Yet the real story is contrast: dark with light, hard with smooth. Jay originally designed the light garden to be viewed from inside the home, but he revised the plan by adding

Mounted on a slender stone pillar, a bronze sculpture by Pavel Efremoff marks the start of the journey into the light garden. Golden sedge (*Carex oshimensis* EverColor 'Everillo') evokes and catches rays of sunlight; spheres appear to roll downhill.

a walking path. Columnar yew plum pines (*Podocarpus macrophyllus* var. *maki*) act as sentries, their dark green-needled foliage casting figurative shadows over mondo grass (*Ophiopogon japonicus*); pools and strips of golden sedge mimic the movement of the sun across the western sky.

Weathered granite and clipped boxwood spheres appear to roll down the hillside toward the creek, suggesting an abandoned game of marbles played by unseen giants. The illusion of movement is created by the careful placement of these spheres as they cross the main axial pathway. In fact, spheres are a repeating element throughout the property, a link that brings a subtle sense of cohesion to the series of otherwise distinct garden rooms. The meandering pathway of hammered hardwood mulch blends with the forest yet facilitates easy strolling, drawing the garden visitor from the industrial sculpture at the entrance to the life-size pirouetting dancer at its end. From here one steps down into a gravel meditation circle, a pared-down space furnished with a simple stone stool, a granite sphere, and "Uaundo," a metal sculpture by local artist Jim Weitzel.

Although the light garden is always beautiful, Jay's favorite time of year in it is in early April when the trees are just leafing out, their fresh green a celebration of new life. Although well traveled by deer, the resident wildlife does little damage here—except for the squirrels, who insist on digging up the mondo grass. For now, Jay goes along with their game and continues to replant it on a regular basis.

Upper Woodland Garden

For this space, Jay was inspired to create a deconstructed room, with doors, walls, floor, and ceiling

OPPOSITE A sculpture of a twirling dancer by Benjamin Parrish signals the terminus of the light garden path.
BELOW The light garden ends with a meditation circle.

ABOVE Antique Chinese doors serve as the entrance to the upper woodland garden, a shady enclave where hostas, trilliums, ferns, and maples thrive.

LEFT In this deconstructed room, a red Chippendale bench fronts an innovative privacy screen embellished with three framed metal leaf forms.

represented by hardscape or plant material. It is not fully enclosed: Jay believes that leaving some elements out engages the mind and allows the visitor to personalize the experience. The Chinese doors through which one enters are around a hundred years old. Originally, they would have been much taller, but skilled craftsmen cut them down and affixed reproduction hardware to reimagine them for modern-day application. Jay applies linseed oil twice a year as a preservative, since although they were intended to be exterior doors, they would most likely have been protected by an overhanging roof. To build something similar in this garden would have been possible but seemed too stylized; Jay opted for a simplified door frame instead, allowing one's imagination to add any other details, exotic or otherwise.

A short path leads up to the distinctive red Chippendale bench, whose elevated position suggests power. In fact, Jay jokes that it's a great place to sit if you've had a bad day, as it makes you feel in control as you look down on the surrounding magical kingdom. Behind it is a triptych of gray cement board walls, an unexpected industrial design element that serves to screen out a neighboring home; the center panel projects forward, adding dimension.

Planting this area had its challenges, not least the deep shade and poison ivy that continues to invade from the adjacent property. Deer also enter this space from above, thanks in part to neighbors who feed them. While they have done little damage to the plants (Jay sprays any hostas and coral bells that may be nibbled), they have broken some of the glass art in this area after being startled.

This garden peaks first in spring, with the coppery new growth of autumn ferns and emerging hosta foliage, and then again in fall, when the fiery colors of Japanese maples create a strong visual connection to the hardscape. Yet as with all the chapters Jay has created in this garden, there is something to explore and experience at any time of year.

Top 10 Plants

ORANGE ROCKET BARBERRY (*Berberis thunbergii* 'Orange Rocket'). Vibrant orange new growth matures to burgundy then turns brilliant crimson in fall. This easy-care cultivar is ideal for containers and landscapes, with a more upright form than many; 4 feet tall, 2–4 feet wide. Drought tolerant once established. Full sun, zones 4–9. Deer resistance: A.

GOLDEN SPIRIT SMOKEBUSH (*Cotinus coggygria* 'Golden Spirit'). Brilliant chartreuse foliage glows in the garden from spring until fall, when it turns fiery shades of red and orange. Deer may nibble the emerging new growth but rarely do major damage. Reaches 8 feet tall, 6 feet wide; for the best foliage display, although sacrificing the "smoke" (flowers), prune hard to 2 feet in spring. Drought tolerant once established. Full sun–partial sun, zones 4–8. Deer resistance: B.

CASCADE FALLS BALD CYPRESS (*Taxodium distichum* 'Cascade Falls').
Tolerant of wet soils, this compact weeping variety is typically grafted onto an upright rootstock and then allowed to cascade. Grows slowly to about 10 feet tall in as many years; it may eventually reach almost double that. The feathery green foliage turns orange in fall. Full sun, zones 4–9. Deer resistance: C.

SKYLANDS ORIENTAL SPRUCE (*Picea orientalis* 'Skylands').
Late winter and spring needle color is a brilliant yellow, which highlights the small red, male pollen cones and larger purple female cones, resulting in a spectacular display. Grows 30 feet tall, 10–12 feet wide. Full sun–partial sun, but to prevent sun scorch in hotter climates this golden pyramidal conifer is best grown in dappled afternoon shade. Zones 4–8. Deer resistance: B.

AUTUMN FERN (*Dryopteris erythrosora*).
This evergreen fern, 2.5 feet tall and wide, thrives in moisture-retentive soil. New fronds are often copper-colored, with a more pronounced hue with increased sunlight in late summer and fall. Full shade–partial sun, zones 5–9. Deer resistance: A.

RHODE ISLAND RED JAPANESE MAPLE (*Acer palmatum* 'Rhode Island Red').
An outstanding, upright dwarf Japanese maple, 6 feet tall and wide, with burgundy foliage that assumes shades of orange and crimson in fall. Partial sun–partial shade, zones 5–9. Deer resistance: B.

FLORIDA SUNSHINE YELLOW ANISE (*Illicium parviflorum* 'Florida Sunshine'). An easy-care, evergreen shrub, 5 feet tall and 3 feet wide, that thrives in well-drained, moisture-retentive soil. The small brown fruit smell of aniseed when crushed. Partial shade–partial sun, zones 6–9. Deer resistance: A.

FEELIN' BLUE DEODAR CEDAR (*Cedrus deodara* 'Feelin' Blue'). This prostrate, spreading conifer reaches 2 feet tall and 6–8 feet wide, taller if trained as a small weeping standard. It prefers full sun and well-drained, moisture-retentive soil but is drought tolerant once established. Damage by deer appears to be primarily as a result of rutting rather than browsing. Zones 6–9. Deer resistance: C.

EVERCOLOR EVERILLO SEDGE (*Carex oshimensis* EverColor 'Everillo'). A low-maintenance evergreen sedge that forms a loose fountain of golden foliage, 16 inches tall and 2 feet wide, in moist gardens. Partial shade–full shade, zones 5–11. Deer resistance: A.

GOLDEN PACIFIC SHORE JUNIPER (*Juniperus conferta* 'Golden Pacific'). This slow-growing, ground-hugging, easy-care golden conifer is a workhorse in the garden. Reliably resistant to deer and rabbits in Jay's experience, although Rutgers lists it as being occasionally severely damaged. Grows 1 foot tall, 5 feet wide. Full sun–partial sun, zones 6–9. Deer resistance: C.

A Garden of Survivors

where serendipity and inspired design go hand in hand

The multi-tiered Italianate fountain was almost a deal breaker, seemingly incongruous with the casual vacation home Janet and Joel were seeking. Yet the sandy, unpaved road that led to the home in the Grand Mere neighborhood of Michigan piqued their curiosity, so putting their initial misgivings aside Janet and Joel arranged to view the property. Parking adjacent to an unpretentious, two-story building they began to relax, satisfied that this was going to be just right for their needs after all. Smiling, the realtor explained that this was just the guest house; the primary home was farther up the hill.

The main residence, built around 1916, was a quirky blend of architectural styles with an intriguing history. The basement still houses a small room that appears to have been used as a safe room or prison cell. The 18-inch-thick concrete walls, ceiling, and floor that encase

QUICK FACTS
LOCATION: Stevensville, Michigan (zone 6)
SOIL TYPE: sand
PROPERTY SIZE: 2.5 acres
PROBLEM CRITTERS: deer, rabbits, groundhogs, voles
OTHER CHALLENGES: dry shade (cast by native pines); exposure to strong, sustained winds; lake spray during violent storms

DESIGN CREDITS:
home of Joel Guillory and Janet V. Burch
designed by Maria Smithburg, Artemisia Landscape Design

the space—together with three sets of steel doors complete with viewing grille—certainly suggest the latter, and it is believed that the home was originally built for one of Al Capone's infamous gang members. For Janet and Joel, the primary consideration was the location, however. They both worked long hours in Chicago, Janet as an ophthalmologist and Joel as a cardiac surgeon, and they relished being able to escape the city on weekends. "It felt as though we were thousands of miles away," Janet recalls of that first visit. The setting on a sandy dune above Lake Michigan, together with a private dock on an inland lake, afforded ample opportunities for bird watching and hiking, two of their favorite pastimes.

It took a few years to update the home, a priority being to open up lake views to the north, but when it was time to tackle the overgrown landscape Janet was at a loss. The existing garden was reminiscent of a neglected public park: an asphalt parking lot installed by the previous owners dominated the sunniest area, the shady entry garden linking the guest house and main home was mostly weeds (accented by the ornate fountain), and the large, weedy lawn was bordered by an unkempt bank of yet more weeds. Wild grapevines grew up many of the trees, and poison ivy was rampant, a real problem for Joel and Janet, who are both highly allergic to this pernicious weed. They had to don special protective outfits that covered them from head to toe before venturing outside to tackle the noxious thug. "It was insane!" Janet laughs.

Although Janet had enjoyed growing vegetables as a child, encouraged and mentored by both her parents, gardening in New Orleans was a far cry from Michigan. Used to hot, sultry summers that supported the year-round exuberant growth of colorful tropical plants, she was now faced with long, snowy winters and sustained winds of fifty miles per hour that sprayed water from the lake across the garden and onto the windows, conditions that killed all but the hardiest of plants. "The winter storms make it hard to be outside, let alone do any gardening," she says. It is not unusual for shallow-rooted conifers to be torn from the ground and branches to be scattered across the garden. Late freezes are also a challenge, especially to her favorite hydrangeas, which can suffer repeated dieback in the unpredictable spring weather.

This firsthand experience has taught Janet that regardless of what books and local experts might

LEFT A trail leads to the private dock, a favorite spot to sit and watch the many marsh birds.

ABOVE Songbirds can often be seen perching on the cattails or wetland shrubs that surround the dock.

recommend, all her planting and transplanting must be done in spring, as soon as the ground can be worked. Nothing planted in fall—whether a perennial, shrub, or tree—survives the winter in this harsh environment.

Taming the Wilderness

Janet realized she needed professional help and a focused vision, so turned to Maria Smithburg of Artemisia Landscape Design. Maria is an intuitive artist, seeing the terrain as her canvas and the plants as her palette. After walking the property and assessing views from each window in the house, Maria took her design inspiration from the lake, which dominated every vista. Using graceful curves and sweeping, parallel arcs, she mimicked the ripples of water when designing steps, paths, and planting beds, these ripples becoming bolder and wider as they moved farther away from the house. The garden is designed to be appreciated from above; every window now offers a framed view of the garden and lake to the northwest or a treetop view to the east.

The asphalt parking area was removed to make way for a series of sunny garden borders and lawn intersected with gently curving paths of Wisconsin Lannon stone; the driveway was reworked to culminate in a wide turning circle and covered carport. Janet also requested a small area to grow herbs; the designer, therefore, included a boxwood-trimmed garden laid out in classic four-quadrant style, its gravel and cobble pathways edged with brick. Pink and white rugosa roses (*Rosa rugosa*) line the nearby bank and perfume the air, and the self-seeding of rose campion (*Lychnis coronaria*) and stonecrops (*Sedum*) into the gravel pathway is encouraged.

A wooden shed and pergola were added at a high point on the bluff; the shaded canopy is now a favorite spot to enjoy lunch or an evening glass of wine while listening to the gentle lapping of the waves

Sweeping curves of lawn, stone, and plants simulate rippling water.

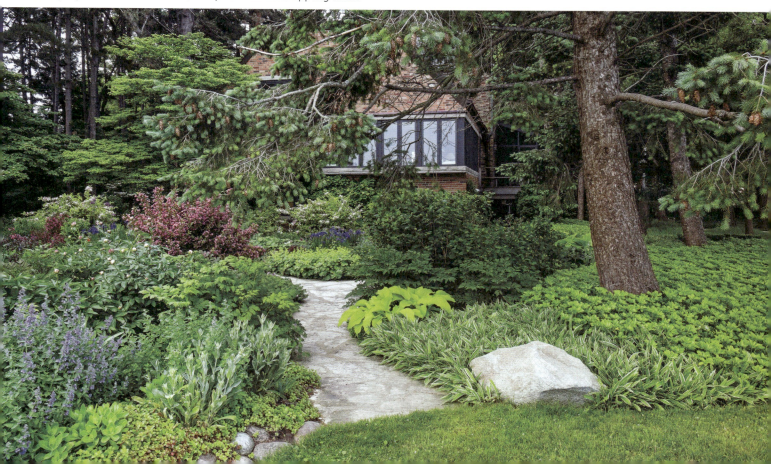

Climbing hydrangea (*Hydrangea anomala* subsp. *petiolaris*), a signature vine used throughout the property, scrambles up the sturdy pergola, offering shade. Deer nibble the rugosa roses (*Rosa rugosa*) but leave enough to be enjoyed.

below. Initially, Janet planted a native grapevine on the pergola but sadly it succumbed to phylloxera. It has been a matter of trial and error to find a replacement vine that will thrive in that location, but a climbing hydrangea (*Hydrangea anomala* subsp. *petiolaris*) is now doing well, while wisteria, clematis, and a kiwi vine (*Actinidia kolomikta*) are still struggling to become established.

Janet next turned her attention to the fountain garden, which she envisioned as an area to showcase shade-tolerant perennials that would also suppress the weeds. Maria suggested first balancing the scale of the towering conifers and oaks by creating an intermediate understory of doublefile viburnum (*Viburnum plicatum* f. *tomentosum*) pruned to resemble small trees, revealing their sculptural form. Maria also addressed a significant design challenge: the formal fountain was not on axis with the main path to the house. To balance this awkward asymmetry, the designer added a Wisconsin Lannon stone and brick patio to span both sides of the path, encompassing the fountain and creating two intimate seating areas, accessed from the home by a series of semicircular steps.

Working with Wildlife

Deer were the least of Janet's concerns in the early days. She naively considered them "quite cute," blissfully unaware of the havoc they could wreak, so her initial plant selection did not focus on deer resistance. It didn't take long, however, before she realized these deer were not merely passing; rather they saw her garden as offering tasty supplements

Hostas are sprayed regularly with a deer repellent, a measure Janet feels is necessary since their bold foliage is a key component of the fountain garden.

to their more traditional diet. That wasn't good news for the countless hydrangeas and hostas Janet had already planted. Her goal, however, was not to create a garden that deer wouldn't destroy; it was to create the garden she really wanted—and then deal with the deer. With this patient approach, Janet fashioned cylinders from chicken wire to protect the young oakleaf hydrangeas (Hydrangea quercifolia) that she was determined to grow, noting that when the shrubs got larger the deer didn't seem to be as interested (although she admits that could easily change in the future). And while the deer ignored the sumptuous hostas for many years, as soon as Janet noticed the emerging perennials were being browsed, she asked her garden maintenance company to begin a spraying regime of deer repellent (used at a slightly higher concentration than usually recommended), which has proven effective.

Rabbits and groundhogs were also increasingly problematic. Emerging daylilies (Hemerocallis), lilies (Lilium), and coneflowers (Echinacea) became choice treats for the groundhogs, the tender young shoots proving particularly irresistible. Janet set to work with smaller chicken-wire cylinders to protect these, which worked well until the lilies grew taller than the cages, at which point the deer ate the exposed buds. The revised plant palette was the result of natural selection, as browsing, soil, and climate all took their toll. "It's a garden of survivors," says Janet philosophically.

From Parking Lot to Painting

Looking at the garden today it is hard to imagine this painterly scene was once a weed-infested parking lot. Climbing hydrangeas thrive on the home's brick façade, while the vibrant foliage of barberries (Berberis) and weigelas ensure color even when flowers are not in bloom. Vertical spikes of Siberian iris (Iris sibirica), bearded iris, and ornamental onions (Allium) punctuate the mounding peonies, catmints (Nepeta), and geraniums, while lady's mantle (Alchemilla mollis) adds a frothy chartreuse ruffle along the paths. Focal points are established using bold foliage such as edible rhubarb (Rheum rhabarbarum), large urns, and a distinctive stone sculpture in the shape of an egg, which appealed to Janet for its simplicity of form.

The dramatic, dry-stacked stone egg was created by local artist Todd Brooks, in the style of one designed by British sculptor Andy Goldsworthy. The sculpture took three weeks to build and necessitated a 4-foot-deep foundation of poured concrete, a pier, and rebar, which runs through both the pier and stonework. The egg itself is constructed from nine tons of Fond du Lac flagstone. Todd first sketched out every layer of stone, meticulously calculating the size of each piece, which was then cut to shape and fine-tuned with a hammer and chisel.

The sculptural dry-stacked stone egg sits at the end of a grass peninsula, framed by flowers and foliage. A nearby bench offers a peaceful setting to enjoy the view across Lake Michigan.

A Garden of Survivors 101

TOP LEFT Pale pink blooms of a volunteer columbine wind through Adam's needle (*Yucca filamentosa*), a cold-hardy species that has naturalized in Michigan. The dark foliage of coral bells (*Heuchera villosa* 'Palace Purple') and weigela (*Weigela florida* 'Wine and Roses') add depth to the scene.

TOP RIGHT Cobalt-blue Siberian iris (*Iris* 'Caesar's Brother') is a vivid counterpoint to the softer-hued lady's mantle (*Alchemilla mollis*) and variegated weigela (*Weigela florida* 'Variegata').

ABOVE The hedged herb garden adjoins this textural feast, the flavors of basil and oregano just moments from the kitchen.

LEFT Edible chives (*Allium schoenoprasum*) and yellow loosestrife (*Lysimachia punctata*) mingle with oxeye daisies (*Leucanthemum vulgare*) on the sandy bank.

Janet insists that the plant selection has been driven more by what survived rather than specifically aiming for deer resistance at the outset. While both she and Maria expected silver-leaved plants to thrive in the fast-draining soil and sunny site, the humidity quickly caused their demise, while the rabbits' voracious appetite for coneflowers thwarted her desire for a vast swath of these classic perennials. Yet even with these restrictions, together they have created many striking vignettes and colorful plant combinations that withstand both the wildlife and weather. Irises have proven to be reliably deer-resistant in the garden, so Janet has enjoyed incorporating many different species and colors.

With so many perennials in the garden, regular division is part of the ongoing maintenance, but Janet puts these divisions to good use by adding them to the informal bank of flowers, creating what has gradually evolved into a botanical Impressionist painting. Entirely spontaneous and serendipitous, this charming scene dances in the slightest breeze, attracting butterflies and bees.

Increasing shade under the pine trees means a change of pace, a reliance on textural foliage and a soothing monochromatic color scheme using shades of green ranging from bright chartreuse to deepest emerald. It is in this area that the designer's artistic flair really shines as she has used bands of different groundcovers, discreetly separated by landscape edging, to create a curvilinear design. Rosettes of pachysandra (*Pachysandra terminalis*) are juxtaposed with wide green-and-white blades of creeping broadleaf sedge (*Carex siderosticha* 'Variegata') transitioning with ease into the clipped boxwood hedge surrounding the herb garden. As more light penetrates the canopy, the larger leaves of hosta and bugbane (*Actaea simplex*) are introduced, with golden Japanese forest grass (*Hakonechloa macra* 'Aureola') cascading at intervals onto the flagstone path.

Transforming the Fountain Garden

Having learned so much from Maria, Janet felt ready to design the plantings of the fountain garden on her own, and it is this area that she feels best reflects her style. Following Maria's cue to first create an understory of shrubs that resemble small trees, Janet added additional doublefile viburnums, as well as introducing oakleaf, smooth, and mophead hydrangeas.

Using the fountain as a focal point and the existing mature rhododendrons as inspiration, Janet began to weave shade-loving perennials into a richly textured carpet, experimenting to see what would survive, and while some of the combinations were planned, she confesses that many are purely by chance. Around the sunken patio, feathery ferns and astilbes jostle with multi-hued hostas and leopard plant (*Ligularia dentata* 'Othello'); the placement of the leopard plant is ingenious, as its burgundy stems and leaf undersides are most easily appreciated when viewed from below. The strategic use of deer repellent on vulnerable plants enables her to manage what is otherwise a low-maintenance design, rich in textural details and elegant in its monochromatic simplicity. "I don't even cut it back in fall," says Janet. "It dies back over the winter and can all be raked up in early spring."

In the fountain garden, the horizontal tiers of a doublefile viburnum (*Viburnum plicatum* f. *tomentosum* 'Mariesii') are allowed to mound to the ground.

A metallic ribbon of Japanese painted fern (*Athyrium niponicum* var. *pictum* 'Pewter Lace') weaves through a carpet of leopard plant (*Ligularia dentata* 'Othello'), assorted hostas, and delicate star astilbe (*Astilbe simplicifolia* 'Hennie Graafland').

Variegated Solomon's seal (*Polygonatum odoratum* var. *pluriflorum* 'Variegatum'), mayapple (*Podophyllum peltatum*), and sensitive fern (*Onoclea sensibilis*) give way to a carpet of sweet woodruff (*Galium odoratum*) at the garden's perimeter.

Survivors: not only the fountain and rhododendrons but cottonwood (*Populus*) snags have been left in place, reminiscent of ancient columns around a sunken amphitheater.

Although Janet was unable to replicate the flamboyance of her New Orleans garden, she did discover several shade-loving perennials with large, tropical-looking leaves and has incorporated many of these into the fountain garden, including Rodgers' flower (*Rodgersia podophylla*) and giant Japanese butterbur (*Petasites japonicus* var. *giganteus*); all have proven to be deer-resistant. To transition from this intricately woven tapestry to the far reaches of the garden, which are more naturalized, Janet gradually simplified the plant selection, using fewer and fewer plants, until only sweet woodruff (*Galium odoratum*) and lily-of-the-valley (*Convallaria majalis*) remain.

Janet laughs as she recognizes the irony, admitting that the fountain garden is now one of her favorite areas.

Top 10 Plants

BLACK BUGBANE (*Actaea simplex* Atropurpurea Group). A wonderful architectural perennial, 4–6 feet tall, 2–4 feet wide, with dark, divided foliage and numerous tall spires of fragrant white flowers in late summer. Prefers rich, moisture-retentive soil and protection from strong winds. Partial shade–full shade, zones 3–8. Deer resistance: B.

JOHNSON'S BLUE GERANIUM (*Geranium* 'Johnson's Blue'). Large periwinkle-blue flowers are held high above the finely cut foliage in late spring. In fall the foliage takes on brilliant orange and red tones, extending the season of interest even further. To 1–1.5 feet tall, 2–2.5 feet wide; shearing this herbaceous perennial after blooming will help maintain a more compact shape. Full sun–partial shade, zones 4–8. Deer resistance: C.

SENSITIVE FERN (*Onoclea sensibilis*). This popular fern gets its name due to its sensitivity to cold temperatures, the soft green fronds turning yellow and dying down with the first frost. It spreads freely in moist soil and is an effective presence from spring until fall, at 3–4 feet tall and wide. Partial shade–full shade, zones 4–8. Deer resistance: A.

LADY'S MANTLE (*Alchemilla mollis*). Cultivated since the nineteenth century, this classic cottage garden perennial is an attractive edging for borders and paths. The hairy, scallop-edged, light green leaves retain beads of moisture that sparkle like diamonds after rain or the morning dew, and the sprays of lime blooms are popular in floral arrangements. Deadheading is recommended to avoid self-seeding, which can be excessive. Grows 1–1.5 feet tall, 1.5–2.5 feet wide. Full sun–partial shade, zones 3–8. Deer resistance: B.

OTHELLO LEOPARD PLANT (*Ligularia dentata* 'Othello'). Large, leathery, round leaves are olive-green on the upper surface and deep burgundy on the reverse, a perfect foil for the bright orange daisy-like flowers that appear above them in summer. Grows 2–3 feet tall, 1.5–2.5 feet wide. Protect from afternoon sun and provide plenty of moisture for this perennial to thrive. Bait for slugs as necessary. Partial shade–full shade, zones 3–8. Deer resistance: A.

CUSHION SPURGE (*Euphorbia polychroma*). The pale green foliage forms a dome-shaped cushion, 1–1.5 feet tall and wide, topped in spring with electric-yellow bracts. Thriving in dry, well-drained soils and full sun, it is a long-lived perennial, although self-seeding may be a problem if not deadheaded. Zones 4–8. *Warning*: the sap may cause irritation; protect skin and eyes. Deer resistance: A.

VARIEGATED WEIGELA (*Weigela florida* 'Variegata'). Even when not in bloom, creamy-margined leaves of this deciduous shrub shine as if backlit. Tubular pink blooms appear in early summer, attracting hummingbirds. An easy-care specimen shrub for the landscape, 6–8 feet tall and wide. Full sun, zones 4–8. Deer resistance: B.

SPILLED WINE WEIGELA (*Weigela florida* 'Spilled Wine'). This compact variety, 2–3 feet tall and wide, maintains a tidy mounding habit even without pruning. The dark foliage retains its color well throughout the season and offers an exciting contrast to the deep pink tubular flowers that appear in late spring. Full sun, zones 4–8. Deer resistance: B.

PLUME POPPY (*Macleaya cordata*). A vigorous, spreading and self-seeding perennial that some may consider aggressive to the point of being invasive, so seek advice from your local extension office before planting. Grows 5–8 feet tall, 2–4 feet wide. The lobed leaves are highly attractive, especially when backlit by the sun; the airy flower panicles are reminiscent of smoke plumes, hence the common name. Full sun–partial shade, zones 3–8. *Warning*: the sap is toxic; wear protective gloves when handling. Deer resistance: B.

YELLOW BARRENWORT (*Epimedium ×versicolor* 'Sulphureum'). A vigorous groundcover, 1 foot tall, 1–1.5 feet wide, for dry shade with year-round interest. The heart-shaped, evergreen foliage may be tinged with red or bronze in dappled shade. In spring this hardy perennial produces a constellation of diminutive yellow and white flowers. Partial shade–full shade, zones 5–9. Deer resistance: A.

A Garden of Connections
between people, flora, and fauna

In 2003, when Maryellen and her husband Bill set out to find a property near downtown Portland with "a bit more space" for their two young children to grow up in, they had a modest five-acre parcel in mind. Yet when they were shown Westwind Farm, a midcentury modern home designed by Pietro Belluschi, with territorial views across the Tualatin Valley to the Coast Range, the fact that it sat on forty acres of grassland and blackberries didn't deter them. Within half an hour, Bill, who had been looking for a place that showed a sensitivity to the land, knew they'd found it. The home appeared to blend into the surrounding landscape, settling unobtrusively into the hillside in such a way that it visually disappeared, a hallmark of Belluschi's designs. Surprisingly, once inside, the light-filled home became a window to the world, the entire south wall offering panoramic views.

QUICK FACTS
LOCATION: Portland, Oregon (zone 8)
SOIL TYPE: clay
PROPERTY SIZE: 80 acres
PROBLEM CRITTERS: deer, elk, rabbits, voles, gophers
OTHER CHALLENGES: very little shade; 800-foot elevation impacts plant hardiness

DESIGN CREDITS:
home of Michael and Maryellen McCulloch
designed by Ann Lovejoy, Beth Holland, Laura Crocket, Eamonn Hughes, John Greenlee

However, the journey from yesterday to the present day has not been painless. Barely a year after moving in, with two children under the age of three, Maryellen received the devastating news that her husband had passed away suddenly of a heart attack while at work. Somehow, she managed to get through the next few years, determined to stay in the home that she and Bill had found. She began to develop a vision for the land, one that was shared by Michael McCulloch, a leading Portland-based architect and friend of the family, who joined her and the children in 2005.

With a renewed sense of purpose, Maryellen and Michael called in their "Dream Team" of Northwest landscape designers: Ann Lovejoy, Beth Holland, and Laura Crockett. Their initial focus was the four sloping acres immediately surrounding the hilltop home, where the McCullochs requested a saltwater pool, several new multi-purpose buildings, and a large patio space. Creating an environment where wildlife could thrive was also important, hence the inclusion of a naturalistic waterfall and pond to support birds, butterflies, and dragonflies, as well as an abundance of perennials to encourage pollinators. Deer and elk were regular visitors, their presence embraced by the homeowners but a reminder to the design team to work with a suite of deer-resistant plants.

Such an extensive project took many permits, big machinery, and over a thousand tons of rock. They finally broke ground in late 2006 and naively invited the Hardy Plant Society for a celebratory opening the following summer. The land was contoured and terraced under the direction of Eamonn Hughes, whose expertise was called upon for the design and installation of the water features. Each basalt boulder, handpicked for its particular color and shape,

OPPOSITE The home's low-profile design and hilltop location afford sweeping views across the valley.

ABOVE All four walls of the yoga pavilion can be slid open, creating an enchanted setting for outdoor concerts. Wide steps around the pool and pavilion can double as event seating, while the careful placement of large boulders suggests pre-existing natural formations.

RIGHT Between the patio and the yoga pavilion large boulders were set into the existing slope to create a waterfall and pond, while also providing planting pockets for an assortment of deer-resistant shrubs, perennials, grasses, and wildflowers.

was painstakingly set to appear as a natural outcropping under Michael's watchful eye. "When rock is set right, it speaks to you," says Michael. However, winter rains turned the construction site into a dangerous mix of mud bath and water-filled ditches, and the less-than-grand opening in July 2007 was a big, sloppy mess of exposed rebar and caution tape everywhere. "It was hard for them to envision it," recalls Maryellen wryly.

A cedar waxwing eyes the juicy fruits of a black chokeberry (*Aronia melanocarpa* 'Nero').

The yoga pavilion was a key focal point of the new landscape, anchoring one corner of the new outdoor living room, and designed by Michael to complement the house. Every component of the timber frame was meticulously crafted over a period of six months before being assembled on site in one day. It is constructed without a single fastener, eschewing modern western practices in favor of traditional Japanese techniques that rely on intricate joinery.

Michael and Maryellen, both strong supporters of the local arts scene in Portland, envisioned using their new outdoor living spaces to host concerts, fundraisers, and other community gatherings. With that in mind, Michael designed a series of shallow steps around the pool to serve as casual amphitheater seating, with a small lawn behind offering additional space to spread out picnic blankets. Coincidentally, the acquisition of a grand piano and subsequent conversion of the basement into a recording studio expanded the role of the home in a similar fashion and marked the evolution of Westwind Farm into Westwind Farm Studio. "Michael is a very sensitive architect. He created everything with people in mind," says Maryellen. Certainly, the landscape reflects his vision for multiple gathering spaces that can accommodate large groups as well as more intimate spots for quiet reflection.

With the hardscape complete it was time to focus on the plantings. Loose, sinuous beds mimic the contours of the rolling hillside and provide a transition from the strong geometric architecture to the more organic flow of the landscape. Michael specifically requested a restrained plant list, although he isn't afraid of bold colors, red being his favorite. To that end, the designers included large swaths of yellow-blooming Jerusalem sage (*Phlomis* russeliana) for structure, silvery lamb's ears (*Stachys byzantina*) as a low-growing, edging perennial, and extensive drifts of bold-colored daylilies. To satisfy Michael's color preference, the fall display includes vines, conifers, and grasses that all turn shades of scarlet, as does the foliage of many of the deciduous trees and shrubs.

It was also important to provide food sources such as berries, seed heads, and nectar for birds, dragonflies, and the many pollinators that call this sanctuary their home. Even the deer and elk are welcome to a portion of the fare, as Maryellen's "more is more" philosophy encourages growing an abundance of every flower. When Maryellen took ownership of the evolving garden, she added more berry-bearing shrubs to supplement the perennials and fifty thousand spring bulbs (although for the latter to succeed she must do battle with voles and gophers).

Lavender fields were added in 2012, fueled in part by the McCullochs' love of community; outgoing Maryellen has always enjoyed lavender festivals and wondered how hard it would be to transform a few acres into tidy, mounded rows of fragrant blooms, thereby providing an excuse to host events that would bring the neighborhood together. The latest phase of the design began a year later, when the McCullochs turned to John Greenlee to devise a plan for the remaining acreage, recognizing the need to connect all these elements to the wider landscape. The area below the home was originally 20 feet high in blackberries, a vestige of earlier clearcut logging. Maryellen envisioned a meadow here, an informal blend of grasses and wildflowers that would naturalize over time, although she knew that creating and maintaining a meadow would be neither a quick nor an easy project.

Indeed, the path to gardening success has not been without its bumps. In the meadows, some things were not persistent, while in the perennial borders, more aggressive plants quickly overwhelmed their more delicate companions. Maryellen discovered the hard way that tickseeds (*Coreopsis*) do not like being covered up with compost ("I killed an entire bed that way"); even some pine trees succumbed to crown root rot after being mulched too deeply. Then there are the plants she should probably never have planted at all, such as bronze fennel (*Foeniculum vulgare* 'Purpureum'), which while beautiful to look at, self-seeds with abandon, and the cattails (*Typha*) she now struggles to eradicate.

Maryellen and Michael see themselves as stewards, not owners of this extraordinary property. They marvel at the intricate web of ecosystems it hosts and are honored to play a small role in its continuation. "We try to understand what the land wants to be," Maryellen says, adding, "I think that the reason we're put on this planet is to learn and

Caterpillars of the cinnabar moth feast on the flowers of tansy ragwort (*Senecio jacobaea*), devouring the blooms before this noxious weed can set seed.

A Garden of Connections 113

"Window to the Gone World" by Lee Kelly invites contemplation and interpretation.

grow. I hope I always continue to do that." She takes that learning seriously, noting that observation is key to understanding life cycles. Each week she takes extensive notes of necessary land management, recording bloom times or problems for future reference. For example, they had never had a problem with tansy ragwort (*Senecio jacobaea*), a common invasive weed of disturbed meadows, until the Oregon Department of Agriculture sprayed the Portland area with a biological pesticide to eradicate gypsy moth caterpillar in 2016. Suddenly the weed became a significant pest. The following year Maryellen noticed dozens of cinnabar moth caterpillars covering the ragwort. Research revealed that this noxious weed is their primary food source and that the caterpillars' presence indicated a biological control was taking care of the weed, without human intervention.

For the next chapter, Maryellen hopes the boundaries between meadow and perennial border are increasingly blurred and the overall aesthetic becomes even more naturalistic. As the sun goes down, the garden takes on a magical quality, the night sky reflected in the mirror-like surface of the pool. The sounds change from birdsong to the quiet hum of evening insects and a few noisy frogs. A sense of peace settles over the land like a sigh. It is a legacy, held in trust by Michael and Maryellen for future generations. It is their gift to each other, to their community, and to nature.

Perennial Borders

A series of delightfully floriferous perennial borders surrounds the home, patio, pool, and yoga pavilion. Such casual plantings can quickly become a weedy jumble without considerable planning and ongoing editing. To avoid this, Maryellen has planted both oxeye daisies (*Leucanthemum vulgare*) and larger-bloomed Shasta daisies (*L.* ×*superbum*) in all the borders as well as the meadows, lending a sense

TOP LEFT Beebalm (*Monarda* 'Raspberry Wine') mingles with native white yarrow (*Achillea millefolium*).

TOP RIGHT Nibbled but not devoured, annual nasturtiums (*Tropaeolum*) bloom so prolifically, the damage is rarely fatal.

ABOVE Shasta daisies (*Leucanthemum* ×*superbum*), seen here as a backdrop to sea holly (*Eryngium giganteum* 'Silver Ghost'), are a unifying perennial throughout the garden.

RIGHT Daylilies are safe from passing deer, who prefer to bypass this narrow, stepped pathway in favor of easier routes.

of unity. These have varying companion plants in the different locations, but the repetition establishes a feeling of order and calm.

Observation of the habits of the resident herd of five females and two or three males that roam freely on the land was essential in achieving an acceptable level of deer resistance. Russian sage (*Perovskia atriplicifolia*) and three varieties of reliably deer-resistant aromatic beebalms were used in the more heavily trafficked parts of the garden, and Maryellen has observed that spiky plants such as yucca and sea holly are rarely bothered, so incorporates those in susceptible areas also.

While daylilies are occasionally severely damaged by browsing deer, they thrive in large swaths here. Maryellen attributes their success to careful placement—primarily flanking a steep flight of stone steps that the deer seem to avoid. Growing more reliably deer-resistant grasses and spiky sea holly at the flight's landings may also dissuade the deer from exploring it. She has noted, too, that the deer appear to prefer walking on soft earth or grass—not on hard concrete.

Other deer-management strategies include fencing young trees to avoid bark damage and planting most things in vast quantities. "If I plant fifty of something, the deer can have three or four," Maryellen says. Nasturtiums and other vigorous annuals quickly recover from browsing, and she even manages to grow a few roses by selecting tall varieties and tethering them to fences. Although the lowest blooms are eaten, the bush as a whole survives, and the highest flowers remain unscathed.

Maryellen grew elephant garlic (*Allium ampeloprasum*) one year and noticed that it was completely ignored by the deer, prompting her to try more plants from the onion family. Now ornamental onions are one of her favorite bulbs in the garden; she enjoys Purple Sensation (*A.* 'Purple Sensation') in several different areas. No repellent sprays are used nor are tall fences erected (with the exception of the vegetable garden). Maryellen isn't trying to repel or exclude the deer but seeks a balance so they can all co-exist harmoniously. "It is absolutely thrilling to see them so close to us—one of the true gifts and treasures of this place." If the deer or elk eat some of her plants, she doesn't get angry but blames herself, for placing something where they could find it!

During the year the perennial borders evolve through several different color schemes. In spring, purple ornamental onions and columbines punctuate a sea of yellow daffodils, while midsummer sees mainly warm shades of gold, red, and purple, before the fiery autumnal display rounds out the gardening calendar. Traditional perennial gardens can appear barren during winter dormancy, so the addition of both evergreen and deciduous trees and shrubs ensures four-season interest. One of her most successful shrubs is Gro-Low fragrant sumac (*Rhus aromatica* 'Gro-Low'), a robust groundcover with foliage that is fragrant when crushed and tinged pink in summer before turning vibrant shades of red in fall.

Siberian cypress (*Microbiota decussata*) is a favorite evergreen conifer, with layered feathery foliage that transitions from bright green in summer to purple-red during fall and winter. Its placement at the base of the towering, vine-covered stone fireplace helps to balance the scale of the masonry, which might otherwise be overpowering. Michael especially loves this vignette in fall when the Virginia creeper (*Parthenocissus quinquefolia*) turns deep red, echoed by the adjacent flame grass (*Miscanthus* 'Purpurascens').

Meadow

Grass ecology expert John Greenlee designed the master plan that connected the home, perennial beds, and lavender fields to the larger landscape. A key to this was how one moved through the space, so he began by modifying the driveways and introducing a network of trails through the existing pastures. John used flags to lay out the trails, introducing "sexy curves" to invite visitors into the space.

Distant views across a meadow are framed by lichen-encrusted branches. Beyond lies an additional forty acres the McCullochs purchased, now protected as a conservation easement.

Mown paths through the meadows are highlighted by sentry-like Italian cypress (*Cupressus sempervirens*).

These paths also direct attention to specific vantage points and vistas, carefully orchestrating the sensory experience. The paths were leveled with an excavator to create 6-foot-wide mowing strips, which were then seeded with native fescue. The paths have transformed the way the McCullochs engage with the land, but looking across the fully grown summer meadow it could be challenging to locate them, so a series of slender Italian cypress (*Cupressus sempervirens*) were planted as markers, adding vertical punctuation points as well as identifying the mown paths. This clever solution almost didn't endure, however; John once watched as a herd of elk reached over the fences that were theoretically protecting the young evergreens. Thankfully most of the trees survived.

John then set about improving the pastures themselves. "I wanted them to look more purposeful," he says. "I wasn't aiming for restoration but rather to accessorize." He did this by adding select deer-resistant plugs (small plants) and bulbs. He also insisted upon the removal of two enormous Douglas firs (*Pseudotsuga menziesii*) that were obstructing the million-dollar view; as he pointed out to Michael, they still had an entire forest of Douglas firs: he could spare two. John introduced the new plants in fall, using flags to mark their location. He told the McCullochs that the new plants would need some space in order to compete with existing grasses, and that even drought-tolerant natives would have to be watered during the first summer. It was also important that they learned to coordinate mowing the pasture with the life cycle of the introduced plants, allowing time for them to grow and set seed.

Keeping the meadow in bloom has been a learning process. Maryellen was less concerned about using only natives, recognizing that many were going to be short-lived; even the purple coneflowers (*Echinacea purpurea*) did not persist beyond a few years. Initially, she sprayed the areas with seed, but the only flowering plants that germinated were red corn poppies (*Papaver rhoeas*). She then tried introducing plugs along the paths and made a point of watering them more vigilantly, but they didn't come back. Two native Pacific Northwest lupins, *Lupinus polyphyllus* and *L. latifolius*, were sown as seed; but after initially doing well, they too are now beginning to fade rather than naturalize as she had hoped, and less than half the Russell Group lupins (*L.* ×*regalis* Russell Group) have returned. Blanketflower (*Gaillardia aristata*) has been more successful after a slow start, and the native white yarrow (*Achillea millefolium*) has done well, growing much taller where it gets watered. The native common fleabane (*Erigeron philadelphicus*) came in on its own, much to Maryellen's delight, and after planting only a couple of yellow evening primroses (*Oenothera fruticosa* 'Fireworks'), she laughs to see that they are now everywhere!

Managing a meadow involves more than planting seed or plugs, however. Many people who have pastures will mow and rake every year, but experience has shown Maryellen that this can kill insect life, as it destroys the habitat before life cycles are complete. She prefers to mow every other year, rotating different sections of meadow, and attributes the noticeable increase in the dragonfly population in part to this strategy. To avoid invasive weeds taking over, she hand-pulls thistles before mowing begins so their seed is not scattered, and also allows the cinnabar moth caterpillars to devour and therefore destroy the ragwort.

John is looking forward to his continuing collaboration with the McCullochs as they endeavor to manage and develop these meadows, and he intends to introduce a lot more bulbs and perennials. "I'm not trying to change the pasture ecology, just adapt it into something more ornamental," he says.

It has taken several years for the meadows to become established, and it will always be an ongoing project that needs careful management.

Lavender Fields

Portland's Mediterranean climate is ideal for growing lavender, but the heavy clay soil is not. Not one to let such a detail dissuade her, Maryellen purchased a thousand cubic yards of a custom soil mix back in 2012 to make her first lavender field of dreams possible, but even that didn't ensure success. They are still experimenting with the best way to minimize weeds, maximize growth, and keep maintenance to a minimum. Using landscape fabric was a problem aesthetically: the lavender was pretty but the paths of black plastic were not. Now they install the fabric directly under the plants, leaving mown grass paths between the rows, a compromise that is also easier to manage.

The McCullochs enjoy sharing Westwind Farm Studio and using it to connect the community to the land. This has now grown into a family affair, with Maryellen's daughter baking delicious lavender shortbread and mixing large batches of lavender-infused lemonade, both popular treats at local lavender festivals. Maryellen also harvests lavender oil and distills it to create a wide range of lotions and creams. "It's a joy to grow," she says with a smile.

Maryellen grows fifteen different lavender cultivars, including white grosso (*Lavandula* ×*intermedia* 'Alba').

Top 10 Plants

JACOB CLINE BEEBALM (*Monarda* 'Jacob Cline'). A hummingbird magnet, thanks to the prolific whorls of scarlet, tubular flowers. Grows 4 feet tall and half as wide. The aromatic dark green leaves resist mildew and are typically ignored by deer. Adaptable to both wet and dry soils. Full sun–partial sun, zones 4–9. Deer resistance: B.

OXEYE DAISY (*Leucanthemum vulgare*). A drought-tolerant perennial, 3 feet tall, 2 feet wide, whose white daisies make excellent cut flowers. Native to temperate Eurasia, widely naturalized in North America. Full sun, zones 3–8. Deer resistance: C.

BLANKETFLOWER (*Gaillardia aristata*). An excellent native cut flower, the flowers display a varying pattern of red, orange, and yellow concentric discs. This colorful perennial, 2 feet tall and wide, attracts bees, butterflies, and birds and is drought tolerant once established. Full sun, zones 3–8. Deer resistance: C.

JERUSALEM SAGE (*Phlomis russeliana*). An upright herbaceous perennial, 3 feet tall and 2 feet wide, noted for its bold silver-green leaves and clustered whorls of lemon-yellow flowers. The seed heads attract birds and make a graphic statement in the border, often persisting into winter. Full sun, zones 5–9. Deer resistance: A.

SIBERIAN CYPRESS (*Microbiota decussata*). A low-growing conifer, 2 feet tall and 8 feet wide, with layered, slightly drooping branches featuring feathery foliage somewhat reminiscent of arborvitae. The bright green summer foliage deepens to purple-red tones in fall and winter. Full sun–partial shade, zones 3–7. Deer resistance: A.

BUENA VISTA LAVENDER (*Lavandula angustifolia* 'Buena Vista'). English lavender is harvested for its oils as well as for culinary purposes. This compact selection, 2 feet tall and wide, has deep blue flowers and gray-green foliage. Full sun, zones 5–8. Deer resistance: A.

GRO-LOW FRAGRANT SUMAC (*Rhus aromatica* 'Gro-Low'). A vigorous, low-maintenance shrub to use as a groundcover, 2 feet tall, 6–8 feet wide. Young growth is tinged with pink; fall color is a vibrant crimson. Leaves and twigs are fragrant when crushed. Full sun–partial shade, zones 3–9. Deer resistance: A.

FIREWORKS EVENING PRIMROSE (*Oenothera fruticosa* 'Fireworks'). Red buds open to fragrant, bright yellow blooms, which close at night. Exceptionally drought tolerant and easy to grow; attracts hummingbirds, bees, and butterflies. Grows 2 feet tall and wide. Full sun, zones 4–8. Deer resistance: B.

KARLEY ROSE FOUNTAIN GRASS (*Pennisetum orientale* 'Karley Rose'). Fat, fountain-like plumes of rose-purple flowers arch over the deep green foliage in late summer, persisting well into fall, when the foliage changes to yellow-tan. Grows 3 feet tall and wide. Full sun, zones 5–8. Deer resistance: A.

SILVER GHOST SEA HOLLY (*Eryngium giganteum* 'Silver Ghost'). Teasel-like steel-blue flowers are surrounded by stiff, ghostly white bracts. A superb architectural selection, 3 feet tall and 1.5 feet wide. Reliably deer-resistant in the McCullochs' hot, dry border, the Rutgers rating notwithstanding. Full sun, zones 4–7. Deer resistance: D.

A Confetti Garden
where pops of color punctuate the landscape

The elegant plant combinations and artistic color echoes that fill this confetti garden can be replicated in the south of England, North Carolina, or warmer regions of the Pacific Northwest. Yet this is Central Texas, where summer temperatures routinely exceed 100°F, the native soil is rocky and shallow, watering is restricted to once a week—and there are deer. Many gardeners despair at the impossibility of creating a beautiful garden when faced with just one of these challenges, yet homeowners Jeff and Diana show that it can be done.

It was love at first sight. The pale limestone home was a mix of Southwest and rustic architectural details, its plentiful shuttered windows promising a light-filled interior. The backyard contained a pool and large cabana where Jeff, who loves to cook, could see himself entertaining family and friends, taking advantage of the long

QUICK FACTS
LOCATION: Austin, Texas (zone 8)
SOIL TYPE: clay, limestone, and caliche
PROPERTY SIZE: 1.5 acres
PROBLEM CRITTERS: deer, rabbits, squirrels
OTHER CHALLENGES: extreme heat and drought; flash floods; watering restrictions

DESIGN CREDITS:
home of Jeff Eller and Diana Kirby
designed by Diana Kirby, Diana's Designs

outdoor season that the climate affords. Meanwhile Diana, a professional landscape designer, saw the lack of any significant landscaping as a bonus, a blank slate on which she could create her dream garden. Just hours after visiting, Jeff and Diana made an offer and prepared to make this their new home.

As with so many properties in Austin, a fence excludes deer from the back garden, but the front remains open, as local regulations prohibit tall fences. Yet Diana was undaunted by the frequent sightings of deer in the neighborhood, especially as she had firsthand experience gardening with them in their previous garden. Her advice to clients has always been to work around the deer's known paths and habits, just as you might with a rambunctious family dog, and so her first strategy to outwitting the deer was to observe them. Diana noted that during spring and summer she would see a small herd several times a week, but that after heavy rains they would visit less often, finding food sources elsewhere.

And deer are only one of the challenges Diana and Jeff faced. The soil is composed of sticky clay, limestone, and caliche, which may appear as small rock-like lumps or as impenetrable layers ranging from several inches to several feet in thickness. Amending the soil is a big part of gardening here, together with creating berms and raised beds to achieve a usable depth of soil. Compounding the problem of the rocky, nonabsorbent soil, severe flash

Set well back from the road and backing onto a greenbelt and canyon, Jeff and Diana's new home offered both privacy and plenty of space to entertain.

floods are not uncommon, causing rain water to rush off the land, washing away mulch and creating deep gullies during a deluge. Not that Austin has a reliably wet climate. Rather, it is a climate of extremes, and during a recent drought, Diana recalls that deer ventured right up to the house, drinking from the fountain and tipping birdfeeders to get to the seed, they were so desperate for food and water.

Selecting plants that thrive in such contradictory weather patterns and inhospitable soil can test even the most experienced gardeners. While grasses, yuccas, and agaves might be planted for reliable drought tolerance, they will rot during a wet winter. In an attempt to solve such riddles, Diana uses her garden as a testing ground for plants, trialing new introductions and experimenting with soil amendments to outwit the deer as well as the weather while saving her clients money and frustration. She uses native and adapted plants in her garden, having discovered that plants indigenous to South Africa, Mexico, and even Japan often flourish in Austin. However, her true focus is on whatever survives.

Overall Vision

For her new home's landscape, Diana wanted both tranquil spots and "a happy garden, with confetti pops of color." She began the transformation in the fenced back garden, where deer did not pose a problem. Besides the cabana and pool, this area initially consisted only of rocks, dirt, and stumps. Here she designed some beds, cut down stumps, put in a playscape for their two young children, and added a small vegetable garden and lawn before declaring the back of the property complete.

She then turned her attention to the large, open front garden. The two curving beds leading to the front door were established, but they were planted with "one of everything." The previous owners were not gardeners; some plants were out of proportion to the scale of the house, others were not going to survive due to incorrect placement, or were just not her style. Diana gradually began changing things and making it her own. Her aim was to have these beds lead the eye and guests to the front of the home, setting the stage for the home's interior as well as the rest of the garden. "I wanted it to be welcoming and inviting—a rainbow of color," she says. Diana also turned the area at the end of the driveway into a charming woodland garden that can be viewed from several rooms within the home; this is the transformation she is most proud of.

Inevitably, the passing of time in a garden means growth and evolution, some welcomed, some not. The previous owners had lined one side of the driveway with several Natchez crapemyrtles (*Lagerstroemia* 'Natchez'), a popular variety valued for its profusion of white blooms and attractive peeling bark. Unfortunately, these have now grown so tall that the blooms are hidden from view, and their beautiful

The presence of deer has not limited the artistry in this garden, where variegated foliage echoes the pale limestone.

ABOVE A cool spring and ample rainfall resulted in an especially vivid spring display of snapdragons (*Antirrhinum majus* 'Sonnet Orange Scarlet').

LEFT The white flowers of the Natchez crapemyrtles (*Lagerstroemia* 'Natchez') are now lost in the overhead canopy, and their mottled caramel bark is increasingly obscured by shrubs.

bark is mostly obscured by a dense hedge of evergreen Texas mountain laurels (*Sophora secundiflora*), which have outgrown their allotted space. Diana acknowledges that she will have to address the problem in the near future, commenting that as much as she loves the exfoliating bark, the need for winter interest is less of an issue in Texas, where the season lasts barely three months, and the location of the crapemyrtles is such that she can rarely see them in winter anyway. "I need to remember that I don't have to keep something simply because it's there." This painful realization and acceptance is one Diana often has to walk her clients through.

Front Garden

A limestone pillar and tiered, serpentine beds mark the entrance to the property; raising the soil level improves drainage and provides greater depth for roots to penetrate since the native soil is notoriously hard to plant in. The color palette was inspired by the cheerful colors of a Talavera birdbath. Early in the season, this bed showcases orange and lemon flavors from narrow-leaved zinnia (*Zinnia angustifolia*), tropical milkweed (*Asclepias curassavica*), and Berlandier's sundrops (*Calylophus berlandieri*), with snapdragons (*Antirrhinum majus* 'Sonnet Orange Scarlet') repeating the orange-scarlet tones of the Mexican pottery. Silver lamb's ears (*Stachys byzantina*) and variegated yucca (*Yucca filamentosa* 'Bright Edge') ensure year-round color, foliage interest, and bold textures to offset the myriad smaller, deciduous leaves within the design.

A bold, blue, whale's tongue agave (*Agave ovatifolia*) establishes a focal point in the upper bed and balances the birdbath. Golden thryallis (*Galphimia gracilis*) and Cuban buttercups (*Turnera ulmifolia*) bring sunshine into the late-season display, while esperanza (*Tecoma* 'Bells of Fire') and autumn sage (*Salvia greggii*) add the "happy colors" Diana is so fond of.

To the right of the driveway, mature trees and shrubs define the property boundary, underplanted with mounds of drought-tolerant hairy Acapulco wedelia (*Wedelia acapulcensis* var. *hispida*) that weave between a spiky Wheeler sotol (*Dasylirion wheeleri*) and a sago palm (*Cycas revoluta*). A cobalt-blue ceramic pot planted with a Mexican lily (*Beschorneria yuccoides* 'Flamingo Glow') establishes a focal point here and connects to the colors of the Talavera pottery opposite.

The long driveway skirts an expansive lawn peppered with Nuttall's oak (*Quercus texana*), post oak (*Q. stellata*), and bur oak (*Q. macrocarpa*), leading to the garage and parking area. A beaked yucca (*Yucca rostrata*), sparkling in the sunshine, indicates the wide meandering path that leads to the home's front entry, while a tall yew plum pine (*Podocarpus macrophyllus*) punctuates the border, its deep green foliage prominent against the home's façade. Unlike the more riotous colors chosen for the back garden and the citrus shades at the property entrance, the borders flanking this path have a softer monochromatic

LEFT The orange and yellow flowers of tropical milkweed (*Asclepias curassavica*) and hairy Acapulco wedelia (*Wedelia acapulcensis* var. *hispida*) contrast brightly with the bold cobalt-blue pot.

BELOW Whale's tongue agave (*Agave ovatifolia*), fascinating boulders, and a colorful Talavera birdbath combine to create an intriguing vignette.

ABOVE Esperanza (*Tecoma stans*) self-seeds in the garden, the dangling seedpods as ornamental as its golden yellow flowers.

OPPOSITE *Zinnia elegans* 'Green Envy', just one of the heat-tolerant zinnias Diana relies upon for summer color.

BELOW An exciting medley of shade-loving foliage plants thrives beneath a pomegranate tree (*Punica granatum*). This protected bed retains moisture much longer than those in direct sun, reducing the need for supplemental water.

scheme in shades of pink, lavender, and purple, brightened with white and chartreuse.

While many Austin gardens focus on drought-tolerant plants with a predominance of spiky, silvery gray foliage, Diana's garden captures the spirit and romance of an exuberant, floriferous English cottage garden but redefines the style for the local climate. Heat-tolerant zinnias add reliable summer color as well as cut flowers for the home, while the intoxicating, spicy scent of Cheddar pinks (*Dianthus gratianopolitanus*) perfumes the spring air. A pink and lemon bicolor lantana (*Lantana camara* 'Mozelle') is edged with lavender-blooming society garlic (*Tulbaghia violacea*); pink skullcap (*Scutellaria suffrutescens*), which is often evergreen in this climate, adds a low-growing pink froth when in bloom.

To the right of the pillared entry, a pomegranate (*Punica granatum*) and loquat (*Eriobotrya japonica*) cast dappled shade and provide delicious fruit.

Beneath these, Turk's cap (*Malvaviscus arboreus* var. *drummondii* 'Pam Puryear'), assorted ferns, Persian shield (*Strobilanthes dyerianus*), Japanese plum yew (*Cephalotaxus harringtonia* 'Prostrata'), and a bold-leaved root beer plant (*Piper auritum*) thrive in the partial shade and add an unexpected tropical touch while continuing the color theme of the sunnier areas. Diana has so far been fortunate in her choice of plants here: "The deer would almost have to ring the doorbell to get to them!"

Where many homeowners might request year-round interest in their front garden, Diana is more open to including an array of herbaceous perennials. Since the winter season is so short and she is outdoors less during those months, she is willing to have some bare spots in the borders at that time of year. She does, however, still include a captivating array of evergreen foliage, including heavenly bamboo (*Nandina domestica* 'Flirt'), Chinese mahonia (*Mahonia fortunei*), and Chinese fringe flower (*Loropetalum chinense* 'Ever Red').

The front border wraps around the home to connect with a 6-foot-tall fence, erected to keep the deer from gaining access to the back garden. Here autumn sage, Mexican bush sage, and esperanza (*Tecoma stans*) explode in colorful abandon for many months, their reliable drought- and deer-tolerance as well as their low-maintenance making them welcome additions to the garden.

The fence's gate, designed by Diana, features a cleverly positioned "nose hole" that allows their two dogs, Fletcher and Max, to be entertained while watching children play and four-legged friends being walked. Cypress vine (*Ipomoea quamoclit*) scrambles over both fence and gate, softening the lines and attracting hummingbirds with its tubular scarlet blooms; it's an annual, but it readily sets seed, coming back with a vengeance each year.

Although plants are selected for their deer resistance, Diana reminds both her clients and herself that no plant is ever truly deer-proof. "Every season, every day is new. So, what they didn't eat last week, they might eat this week—it might be a different deer." To prove that point, just a short drive away, Diana has observed lantana being eaten to the ground by deer in her friend's garden, yet the same shrub remains untouched in hers. Surprisingly she also seems to have primarily does and fawns visit her garden, so while spring and summer browsing is a problem, damage by antlers is not. "Fawns never seem to get the memo that they aren't supposed to eat something," she laughs, "and sometimes they seem to just dig something up for fun!"

Thwarting the deer is an endless challenge, especially as Diana has observed that their keen sense of smell seems to attract them to freshly laid native hardwood mulch. Young or tender plants often need protecting with cages, and she has resorted to tying tumble dryer sheets to plants or sprinkling soap shavings on the soil, hoping the scent will deter the deer. Aromatic plants such as herbs and society garlic have also been pressed into service as barriers in an effort to minimize damage to vulnerable plants, especially those she admits to experimenting with that aren't reliably deer-resistant. "I guess you could say I've tried everything," she says, "and some of the time it works."

The perfect height for inquisitive pups, this custom nose grate allows Fletcher to see what's happening in his neighborhood.

Woodland Garden

At the end of the driveway, a flagstone path set in 3-inch Hill Country Mix river rock leads into a heavily treed area. The live oaks (*Quercus virginiana*) and yaupon holly (*Ilex vomitoria*) have been limbed up to create a woodland garden, where a wide range of shrubs, perennials, and annuals thrive in the cooler conditions.

Spring in the woodland garden is heralded with a cheerful display of golden daffodils and lavender Chinese ground orchids (*Bletilla striata*), both of which are slowly naturalizing. As temperatures rise and these enter dormancy, the backbone of textural foliage plants is evident, from the feathery (*Mahonia* 'Soft Caress') to the holly-like (*M. bealei*). Variegated plants brighten this shady area, among them a white and green abelia whose tag has long since been

RIGHT A rusted metal planter, hung in the open woodland garden, features more bright confetti pops of color.

BELOW Both practical and beautiful, this woodland path also diverts water during heavy rain. It is flanked with an exciting combination of foliage plants—and a shoal of cobalt-blue fish swimming by a clump of seaweed-like sedge (*Carex phyllocephala* 'Sparkler').

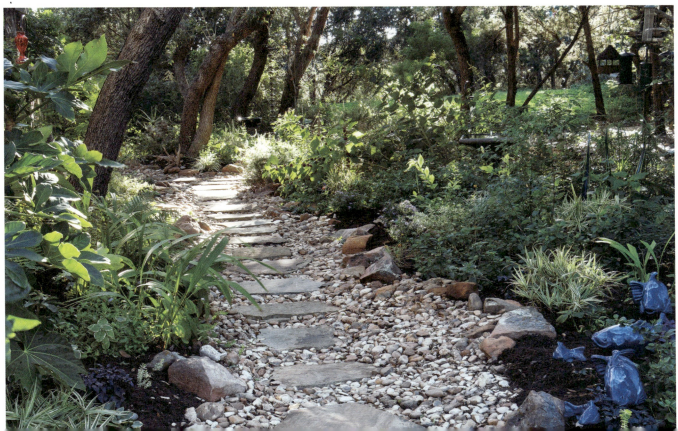

mislaid and a yellow-variegated skyflower (*Duranta erecta* 'Cuban Gold'). Understated details can be lost in brightly lit garden spaces, yet here in the dappled woodland they shine. Twinkling glass stars hang from the trees, catching the light; and a miniature wall planter holds a trailing succulent and a pair of ceramic toadstools.

This garden is alive with the sounds and antics of birds splashing in the bubbling fountain or visiting one of the many birdfeeders; large casement windows ensure these special moments are enjoyed from the breakfast room, kitchen, and family room. Jeff takes care of maintaining the fountain and refilling the feeders, but Diana is in charge of selecting plants that the birds might appreciate. The purple-berried American beautyberry (*Callicarpa americana*) is a favorite with them, together with forsythia sage (*Salvia madrensis*), whose yellow blooms entice not only ruby-throated hummingbirds but bees and butterflies as well. "The garden is my sanctuary, but it is also a sanctuary for the wildlife," she says. "The deer frustrate me—but I love having them here."

LEFT Glass stars bring sparkle and detail to the dappled shade.

BELOW Many different species of birds, including the Carolina chickadee, black-crested titmouse, and northern cardinal, are attracted to the bubbling fountain as well as to the flowers and berries that surround it.

Top 10 Plants

BELLS OF FIRE ESPERANZA (*Tecoma stans* 'Bells of Fire'). A compact selection, 4–6 feet tall, 3–4 feet wide, whose vibrant, fiery red tubular blooms are a siren call to hummingbirds. Unlike the native yellow-flowering species, this variety does not set seed. Full sun, zones 9–11, but Diana has grown it successfully in protected areas of her zone 8 garden. Deer resistance: A.

HAIRY ACAPULCO WEDELIA (*Wedelia acapulcensis* var. *hispida*). A tough native shrub, 2 feet tall, 3 feet wide, that tolerates caliche and drought. It is evergreen in warmer climates but dies to the ground farther north. Golden yellow daisies cover the plant from summer until frost. Full sun, zones 7–9. Deer resistance: A.

CYPRESS VINE (*Ipomoea quamoclit*). Tropical annual vine, 6–15 feet tall, 3–6 feet wide. The fern-like foliage is adorned for many months with scarlet trumpet-shaped blooms that attract hummingbirds and butterflies. It may naturalize by self-seeding in favorable conditions but is easy to control. Tolerates drought but does best with some supplemental water. Full sun, zones 11–12. Deer resistance: A.

MEXICAN FEATHER GRASS (*Nassella tenuissima*). This delightful feathery grass, 2 feet tall and wide, adds movement and light to the landscape. Native to Texas and New Mexico but may be invasive in other regions. Drought tolerant but benefits from water once a week to prevent dormancy. Thrives in well-drained soil. Full sun, zones 6–10. Deer resistance: A.

MEXICAN BUSH SAGE (*Salvia leucantha*). A shrubby evergreen perennial, 2–3 feet tall and wide, popular for its abundant bicolor flowers that are loved by hummingbirds, bees, and butterflies. Drought tolerant but benefits from regular watering. Full sun, zones 8–10. Deer resistance: A.

SOCIETY GARLIC (*Tulbaghia violacea*). These perennial bulbs produce strap-like foliage and clusters of sweetly fragrant, lavender, pink, or white blooms on tall scapes. Both foliage and flowers have a mild garlic flavor and are edible. Plants, 1–2 feet tall and 1 foot wide, are reliably heat tolerant but do best with regular moisture. Full sun, zones 7–10. Deer resistance: A.

PINK SKULLCAP (*Scutellaria suffrutescens*). A low-growing, woody-based herbaceous perennial, 1.5 feet wide and half that in height, featuring snapdragon-like tubular blooms from May through October, atop gray-green, thyme-like foliage. The common name comes from the similarity of the flowers to hats worn by men during the Middle Ages. Full sun, zones 7–9. Deer resistance: A.

JAPANESE ARALIA (*Fatsia japonica*). This tropical-looking evergreen, 6–16 feet tall and wide, is an exciting addition to shade gardens, often tolerating colder conditions than listed. Glossy green leaves clothe the multi-trunked form; drumstick-type cream flowers appear in fall above the leaves, followed by black fruit. Partial shade–full shade, zones 8–10. Deer resistance: B.

AUTUMN SAGE (*Salvia greggii*). A billowing, aromatic shrub, 2–3 feet tall and wide, that thrives in lean, rocky soils. Flowers are borne from spring until frost and may be red, pink, purple, orange, or white. Evergreen in mild climates. Full sun, zones 8–10. Deer resistance: A.

AMERICAN BEAUTYBERRY (*Callicarpa americana*). This native plant prefers forest-floor habitat with organically rich, moisture-retentive soil in partial shade but can adapt to other conditions and fruits most prolifically in full sun. Magenta berries persist after the leaves have fallen in autumn. Grows 3–6 feet tall and wide. Zones 6–10. Deer resistance: A.

A Lake House Garden
where every day is a vacation

J udy and Tom Muntz are a retired couple who have owned a vacation property on the shores of Lake Michigan for over forty years. When they felt the need for a larger home to accommodate their growing family, they immediately thought of their beloved little cottage in the Grand Mere neighborhood, where even today, golf carts are more numerous than cars, and neighbors stop to chat as they stroll down the sandy lanes on their way to the beach. So, when a double lot went up for sale just a stone's throw from their existing cottage, they jumped at the chance to create their perfect lake house.

Working with a local architect and builder, they sketched their ideas out on the back of an envelope. They were adamant that the home should not impose on the landscape, nor block the neighbors' views. The architect exceeded their expectations but suggested modifying

QUICK FACTS
LOCATION: Stevensville, Michigan (zone 6)
SOIL TYPE: sandy loam
PROPERTY SIZE: 0.14 acre
PROBLEM CRITTERS: deer, chipmunks, moles
OTHER CHALLENGES: dry, woodland shade; septic system and underground propane tank on the small lot

DESIGN CREDITS:
home of Tom and Judy Muntz
designed by Anna Brooks, Arcadia Gardens LLC

their initial concept of a single wraparound porch, splitting it into three spacious, independent porches. The final result is a casual light-filled beach house, with clean lines, large windows, high ceilings, and a serene atmosphere that encourages relaxation. With its modestly pitched roofline, the structure seems to settle into the land, suggesting a much smaller footprint than its actual 4,300 square feet.

Before construction even started, Judy knew she needed to find a landscape designer with a sensitive eye and local knowledge of the challenging climate and soils. Arcadia Gardens LLC had reimagined their original cottage landscape many years earlier, so the Muntzes naturally turned to them once again. Designer Anna Brooks led the project, and together with Judy and Tom made an initial assessment to determine which of the existing trees should be kept, and which should be sacrificed to achieve an acceptable balance between allowing light into the home and creating necessary summer shade. Judy's key request was that this garden would require less maintenance than her cottage garden had; above all, she was looking for a garden that reflected the calm ambience of the home she was creating. To achieve this the designer relied on easy-care shrubs and perennials to establish the framework, keeping similar plants grouped together for speedy deadheading and fall cleanup. By using larger sweeps of plants and keeping to a simple color scheme of various shades of green, with white accents and a little plum for depth, she kept a fresh yet peaceful overall aesthetic.

Watering can quickly become a tiresome chore, so the designer suggested installing a zoned irrigation system: the more drought-tolerant natives receive less water, while the thirstier cutting garden and edibles receive more. Providing adequate watering

OPPOSITE Grand Mere State Park on Lake Michigan is popular for its sand dunes and dog-friendly beaches.

RIGHT The interior of the home is light and bright, each window framing a different scene. In only its third year, the garden already looks lush.

BELOW A narrow pathway from the northwest patio leads to the beach, the break in shrubbery facilitating conversation with the neighbors, yet most of the rear of the home is screened from the sandy lane by dense, layered plantings.

on the sandy soil is a significant challenge, as rain or irrigation water percolates so quickly. The designer, therefore, recommended installing drip irrigation hoses directly over the root balls of larger trees and shrubs, since experience has taught her that the more traditional spray heads do not water deeply enough and can be wasteful. Smaller drip irrigation lines were also run to the containers.

To fill in around their trunks would have risked suffocating the hemlocks; a protective tree well was the ornamental solution.

Establishing privacy was the second major priority both for the homeowners and their nearest neighbors. With many of the original shrubs removed, there was now no buffer between the two homes, and since their house took up a significant portion of the lot, Judy and Tom felt exposed. Compounding this sense of exposure was the public beach access path that runs just a few feet beyond the north porch. Tom didn't want beachgoers to feel obliged to wave every time they passed, so the design needed to create screening for privacy but in such a way that still felt neighborly, an important detail in this friendly community. The solution was not a fence but layered plantings; working from inside the home, the designer sited larger trees and shrubs to screen unwanted views of power poles and utilities, making sure the home's large windows framed ever-changing vignettes of foliage and flowers.

Judy is the primary gardener, a gift and interest passed down from her grandparents. She also loves to cook, so requested space in the new landscape design for a few easy edibles as well as a cutting garden. Tom's role is one of helper, mover, digger—and chat-ter. "Tom's idea of gardening is putting his gloves on, standing in the middle of the road, and talking to everyone that comes by," says Judy. He is also a keen bird watcher, a detail that the designer took into consideration: she included many plants that would provide nectar, pollen, seeds, and berries for the birds, bees, and butterflies.

The Muntzes love to entertain their family and friends, so it was important to have outdoor spaces to accommodate gatherings of all sizes as well as an easy indoor–outdoor flow. Each of the three screened porches now opens onto a bluestone patio, and each patio is interconnected by a wide pathway that meanders through the property. There is a notable absence of lawn, a strategic choice on the homeowners' part; they feel the patios make better use of the outdoor space and are easier to care for. "I've never owned a lawnmower and hope to go the distance without one," says Tom.

Major construction projects are rarely without challenges, however, and this was no exception. With a significant change in elevation from one side of the lot to the other, two existing Canadian hemlocks (*Tsuga canadensis*) were in danger of being almost 3 feet below the finished grade. In an effort to preserve as many trees as possible, Arcadia Gardens built a stacked stone retaining wall around the trunks, creating a tree well that has become an interesting focal point adjacent to the northwest patio.

And of course, there is always the challenge of the local wildlife. Deer stroll through the neighborhood daily, and it was not unusual to see one or two cutting through this property when it was still an undeveloped lot. Most of the Muntzes' neighbors choose to spray deer repellent, and those that don't often have plants chewed down to the ground; Judy notes that the gardens subject to the worst damage have no perimeter planting. This is where the designer's firsthand knowledge of deer-resistant landscapes, gathered from her adventures with her own Blue Jeans Garden, has proved invaluable.

Cutting Garden

The northeast corner of the front garden receives the most sun, so this is the perfect location for Judy's cutting garden, where a bounty of flowering shrubs and perennials provides blooms for harvesting. Among the selected plants are some surprising choices for a garden in deer country. The inclusion of roses was non-negotiable, so together Judy and Anna selected several Knock Out roses for their long bloom time and hardiness, as well as some other varieties based on color and their suitability for cutting. Knowing these would be an invitation to the resident herd of deer, the designer screened the roses from view with a dense hedge of deer-resistant trees and shrubs.

Deer-resistant peonies mingle with the roses. White-blooming hydrangeas (*Hydrangea arborescens* 'Annabelle') are sheltered from the harsh afternoon sun and eyes of inquisitive deer by the mature conifers, yet are still within easy reach of floral scissors. Butterfly bushes (*Buddleja davidii* 'Pink Delight') and catmint (*Nepeta* ×*faassenii* 'Kit Kat') add to the profusion of sun-loving flowers for cutting, and the versatile foliage of ferns and hostas can be gathered from the adjacent woodland walk to round out any arrangement. Pink, white, and blue are the primary colors in the cutting garden; Zagreb tickseed (*Coreopsis verticillata* 'Zagreb') is the only yellow bloom that Judy agreed to, persuaded by the selection's abundant petite blooms and feathery foliage.

Judy hopes that by keeping her beloved roses screened and close to the house she will have enough blooms for cutting.

ABOVE Another casual seating area for two under the trees provides close-up viewing of the birds, as does the wide east porch.
OPPOSITE Amethyst Falls Chinese wisteria (*Wisteria frutescens* 'Amethyst Falls') creates a colorful punctuation point as it climbs up a power pole.

Tucked in among the roses and peonies are seasonal herbs and vegetables grown in movable pots that can be repositioned as the pattern of light shifts during the season. This keeps all the thirstiest plants together for easy watering, while the pots themselves can be safely stored away from freezing temperatures during the winter. Within the cutting garden, the bluestone patio features a table flanked by two white Adirondack chaise longues: a perfect spot to arrange an informal garden bouquet or to stretch out and relax with a friend.

Woodland Walk

A wide, dry-laid flagstone path winds from the shaded portion of the front garden to the adjacent woodland walk, which is alive with the sounds of birds. Tom takes great pleasure in refilling the numerous feeders along the walk each day and enjoys watching the antics of red-winged blackbirds, cardinals, and finches.

The oaks, pines, junipers, and hemlocks that form the woodland walk's overhead canopy have been underplanted with a blend of smaller trees and shrubs, both native and ornamental, including Japanese stewartia (*Stewartia japonica*), lilac (*Syringa pubescens* subsp. *patula* 'Miss Kim'), and Ozark witch hazel (*Hamamelis vernalis*). This layered understory buffers the house from the road, providing privacy as well as tentative seclusion from the deer. The perimeter planting was designed to appear natural when it matured, melding to create a dense thicket through which the deer would be unable to pass.

Judy couldn't bring her favorite wisteria from their previous cottage but did ask that one be included in her new garden. Anna selected Amethyst Falls Chinese wisteria (*Wisteria frutescens* 'Amethyst

FAR LEFT Oriental hellebores (*Helleborus orientalis*) can be relied upon for hardiness in Michigan's harsh climate, their winter blooms often lasting until spring.

LEFT Besides bringing depth and drama to the green scene, dark-leaved coral bells (*Heuchera villosa* 'Palace Purple') tolerate a surprising amount of sun—as well as deer pressure.

ABOVE Tiered shrubs and perennials beneath the tall trees create a naturalistic woodland. Ferns, hellebores, and other shade-loving perennials thrive in the filtered light.

LEFT The bold Sum and Substance hostas and bright New Guinea impatiens (*Impatiens hawkeri*) lead the eye down the woodland walk back toward the sunny cutting garden. Neighboring homes are slowly being hidden from view by the maturing perimeter plantings, which also serve to exclude the deer.

Falls') and placed it to scramble up and disguise an ugly but necessary power pole. Tom periodically drives a nail into the wooden pole for the vine to wrap onto, encouraging it to reach skyward—for as far as he can safely reach, at least. Beyond that, Judy says that the wisteria will be left "to do its own thing."

The lowest planting tier comprises large swaths of shade-loving, deer-resistant perennials, including ferns, grasses, oriental hellebore (*Helleborus orientalis*)—and a few dark-leaved coral bells (*Heuchera villosa* 'Palace Purple') to add depth to the otherwise green and white color scheme. Some experts claim Palace Purple is deer-resistant, but results vary across the country; so far, however, this heuchera remains untouched in the Muntzes' garden.

Several large, chartreuse Sum and Substance hostas (*Hosta* 'Sum and Substance') transplanted from the Muntzes' cottage garden now accent the woodland walk. Sentimentality overruled caution here, as Judy admits that the deer might eat these, but she loves this variety enough to spray with a repellent if it becomes a problem. And although the Muntzes specifically wanted to avoid planting large quantities of annuals every year, Judy couldn't resist adding some white New Guinea impatiens (*Impatiens hawkeri*) to line the path, here and elsewhere, loving the splash of brightness. These too are on her "spray if I have to" list.

The meandering pathway curves through the narrow space between the Muntzes' and a neighboring home, where hemlocks and an existing pine help provide screening, before reaching an outdoor shower (perfect for washing off sandy dogs) and a parking area in the back garden.

Back Garden

In the back garden, a generous bluestone patio serves as a grilling station, equipped with a portable barbecue and convenient serving table. Access to the kitchen is via the west porch, whose furnishings include Adirondack rockers, child-sized wicker chairs, and a swing made by Judy's grandfather. This porch and patio are screened from the road by deep beds of existing conifers and new plantings. Once again Judy chose to risk some of her favorite plants, including several hydrangeas (*Hydrangea macrophylla* 'Endless Summer') planted close to the house. In front of these are more reliably deer-resistant plants including astilbes, ferns, and coral bells.

As the sun goes down at the end of a warm summer's day, you are likely to find Judy and Tom pouring a glass of wine and heading to the west porch with Mia, their friendly Airedale mix, trotting at their side. Their home is indeed a peaceful retreat, one that is both welcoming and private, surrounded by a very personal garden. Judy settles onto the heirloom swing with a contented sigh: "I get up each morning, look out, and realize how much I love where we are."

ABOVE Endless Summer hydrangeas are attractive even when not in bloom, their creamy yellow buds echoing the golden Japanese forest grass (*Hakonechloa macra* 'Aureola') and the heart-shaped leaves of a variegated Siberian bugloss (*Brunnera macrophylla* 'Hadspen Cream').

RIGHT Two multi-functional patios at the back of the home are connected by a winding pathway, with deep perimeter plantings ensuring privacy from the road. A burgundy-leaved Japanese maple (*Acer palmatum* 'Bloodgood') contrasts effectively with the American hornbeam (*Carpinus caroliniana*), whose hop-like catkins dangle from the branches.

Top 10 Plants

PEONY (*Paeonia*). The name of the peony shown here is not known, but Festiva Maxima, with fragrant white blooms that have an occasional scarlet streak, would be a good substitute. Even the foliage is a delight for floral arrangements. Judy recommends placing cut blooms in the refrigerator to extend their vase life. Plants are 3–4 feet tall and wide. Full sun, zones 3–8. Deer resistance: A.

GOLDEN JAPANESE FOREST GRASS (*Hakonechloa macra* 'Aureola'). From spring until fall this herbaceous grass creates a cascading waterfall of soft yellow foliage, 1.5 feet tall and wide. Although recommended for partial shade, it can tolerate full sun with regular water. Zones 5–9. Deer resistance: A.

VISIONS IN WHITE CHINESE ASTILBE (*Astilbe chinensis* 'Visions in White'). This robust, compact selection, 2 feet tall and wide, features feathery white plumes on stout, red stems; the greenish flower buds may be browsed by deer, but the foliage is seldom severely damaged. Thrives in dappled shade and moisture-retentive soils, withstands periods of heat and drought. Partial shade–full shade, zones 4–8. Deer resistance: B.

HADSPEN CREAM SIBERIAN BUGLOSS (*Brunnera macrophylla* 'Hadspen Cream'). Pale blue flowers reminiscent of forget-me-nots rise above heart-shaped, cream-margined leaves in spring, on plants 1.5 feet tall and 2 feet wide. Partial shade, zones 3–7. Deer resistance: A.

AMETHYST FALLS CHINESE WISTERIA (*Wisteria frutescens* 'Amethyst Falls').
At 8–10 feet long or more, this twining vine is perfect for smaller spaces. It grows much more slowly than most wisterias, yet lightly fragrant purple flowers appear at a young age. Provide a support for the tendrils to attach to. Partial sun–full sun, zones 5–9. Deer resistance: B.

KIT KAT CATMINT (*Nepeta ×faassenii* 'Kit Kat').
This sterile dwarf cultivar, only 1.5 feet tall and wide, features aromatic foliage and lavender-blue flowers in summer. Reliably drought tolerant and ideal for the front of the perennial border. Full sun–partial shade, zones 3–8. Deer resistance: A.

NORTHERN MAIDENHAIR FERN (*Adiantum pedatum*).
A favorite of florists, this delicate herbaceous fern thrives in damp, shady locations. Wiry black stems support the finely textured fronds. Grows 2.5 feet tall, 1.5 feet wide. Partial shade–full shade, zones 3–8. Deer resistance: A.

AMERICAN HORNBEAM (*Carpinus caroliniana*). This native tree, 20–35 feet tall and wide, offers multiple seasons of interest, with fluted blue-gray bark, hop-like catkins in spring, and outstanding fall color. It is adaptable to a wide variety of soils, from sandy dunes to clay. Full sun–full shade, zones 3–9. Deer resistance: B.

MARIESII DOUBLEFILE VIBURNUM (*Viburnum plicatum* f. *tomentosum* 'Mariesii'). Noted for its distinctive tiered habit, this multi-stemmed deciduous shrub features white lacecap-hydrangea-like flowers in spring and outstanding fall color. Grows 10–12 feet tall, 12–15 feet wide. Full sun–partial shade, zones 5–8. Deer resistance: B.

MISS KIM LILAC (*Syringa pubescens* subsp. *patula* 'Miss Kim'). At 4–7 feet tall and wide, this compact, deciduous shrub is an excellent low-maintenance variety for smaller spaces. It produces an abundance of fragrant lavender flowers in spring. Shows excellent resistance to powdery mildew. Full sun, zones 3–8. Deer resistance: B.

A Collector's Garden

↳ where hostaholics and deer draw a tentative truce

THE GROTTO

BARN QUILT

FOUNDATION BEDS

N

THE DRIVEWAY BED

It seems impossible. How can six hundred juicy hostas and almost three hundred daylilies be ignored by the deer that pass this garden every day? Self-confessed plant addicts Mike and Anita Sheehan have had over forty years to fine-tune their tactics, but then again, the deer weren't always a problem.

When the Sheehans purchased this rural acre in 1971, there were very few deer in this corner of Holland, New York, a quiet town in the foothills of the Allegheny Mountains where dairy farms pepper the pastoral scene. There was also very little house: the builder had run out of money, and they had to knock on their new neighbor's door that first night to ask to use their bathroom, since theirs was far from complete. Once the house was finished, Mike and Anita turned their attention to the garden. When asked about their master plan for the landscape at that

QUICK FACTS
LOCATION: Holland, New York (zone 5)
SOIL TYPE: clay
PROPERTY SIZE: 1 acre
PROBLEM CRITTERS: deer, rabbits
OTHER CHALLENGES: hilly terrain; erosion

DESIGN CREDITS:
home of Mike and Anita Sheehan
designed by Mike and Anita Sheehan

point, they both laugh. "There wasn't one!" says Anita, who Mike concedes is the primary gardener, especially in those early years.

They began with the trees. A local nurseryman advised them to site a tree near the front patio to offer summer shade, so they planted a silver birch (*Betula pendula*), a tree Anita remembers fondly from her childhood. The saucer magnolia (*Magnolia ×soulangeana*) near the front door was a similarly sentimental choice, although sadly it often loses its flower buds to late frosts. Indeed, all plants have to be tough in this snowbelt, where temperatures plummet to minus 20°F and a deep blanket of snow can be expected every winter. Many conifers are good candidates for such situations, however, so when the county offered pine (*Pinus*) and spruce (*Picea*) saplings at a bargain-basement price as part of a reforestation program, they stocked up and planted them atop a berm to provide screening from the road, not knowing then that they would eventually anchor an extensive display garden. Anita explains how yesteryear's trees evolved into today's borders: "We dug a small circle to plant a tree. When the tree grew, we widened the circle—and filled it with plants. We just kept planting more trees, and over time those tree circles joined together." When some of the pines declined, the Sheehans replaced them with silver maples (*Acer saccharinum*) for additional fall color.

Local gardening consultant Sally Cunningham had a more strategic approach to designing the foundation beds. "She arrived one day with a carload of plants—and her ninety-year-old mother," Anita

TOP A barn quilt featuring stylized hosta leaves is a fitting emblem to adorn the Sheehans' home. It is one of several in the area, part of a barn quilt trail to mark the 2018 bicentennial celebrations in Aurora, Wales, and Holland, New York.

ABOVE LEFT Although open to passing deer, this 150-foot-long border showcasing some of the Sheehans' hosta and daylily (*Hemerocallis*) collections, as well as more typically deer-resistant coneflowers (*Echinacea*) and lungworts (*Pulmonaria*), not only survives but thrives thanks to their attentive regime.

LEFT An open area of lawn was important for neighborhood badminton games, initially between the parents of toddlers, later for the growing children.

marvels. "Together we kept moving those plants around until we all liked it." In the back garden, a border known as "the grotto" was dug by their son almost thirty-five years ago, primarily to offer a view from the back patio. Anita promised him at the time that she wouldn't make any more beds. "I haven't!" she exclaims with a twinkle in her eye. "I just expanded the ones I already had!"

The driveway bed is a case in point. Mike points out that it's much more fun than a mowing strip and was a necessity to house all the new hostas and multi-hued daylilies they kept acquiring. Unfortunately, their love affair with these two collectible perennials coincided with an increase in the local deer population. Mike reports that they typically see three deer at a time, but it is not unusual to have eight or nine come through the garden together. With thirty-five acres of woodland behind them and another ten to one side, all interconnected by a series of trails, the deer have easy access onto and through the Sheehans' property. Fencing is cost prohibitive, and since deer consider both hostas and daylilies pure caviar, the Sheehans had a significant but not insurmountable problem. They thwart the deer with a three-pronged approach:

- They plant in vast quantities, so minor damage is barely noticeable.
- They fertilize with Milorganite, which has well-documented (if anecdotal) deer repellent properties.

The foundation beds are anchored by a towering, multi-trunked silver birch (*Betula pendula*), the first tree planted on the property. Later additions include a smokebush (*Cotinus coggygria* 'Golden Spirit') and an assortment of green and gold hostas.

- They spray regularly with Mike's special home-brewed repellent, alternated with Bobbex, a commercial spray.

Both sprays smell for a few hours, but they will not harm or discolor the plants, and they remain effective for approximately three weeks, unless there is a heavy rain, in which case Mike resprays. It takes him just one hour to spray four gallons of his concoction onto susceptible plants.

The Sheehans faced other gardening challenges. Their clay soil was a saturated quagmire during the rainy season, and heavy summer downpours swiftly turned prepared beds into small-scale Erie Canals. The slope into which the home and garden are set presented additional trials. Until they installed a series of drains to divert the runoff, their basement flooded every winter, and any mulch they added was washed away by heavy rain. Anita somersaulted down the hill twice and broke her ankle on another occasion, after impulsively heading up to cut a boutonnière while wearing dress shoes; she now uses garden tools to anchor herself and prefers to work from a kneeling rather than standing position for better balance and safety. Yet despite their advancing years, this spirited couple manage the garden

> **MIKE'S DEER REPELLENT**
> To 1 gallon of water add:
> - 2 TABLESPOONS VEGETABLE OIL
> - 2 TABLESPOONS LIQUID DISH SOAP
> - 2 CUPS HOT SAUCE
> - 2 BEATEN EGGS MIXED WITH 2 CUPS OF MILK (STRAINED THROUGH A SIEVE)
> - 1 TEASPOON GARLIC POWDER
>
> Mix well and use solution immediately, spraying on plants when no rain is expected within 24–48 hours. Rinse sprayer thoroughly when finished to prevent clogging.

The steep slope presents several challenges, including soil erosion.

Anita's birch commemorates some of the events they have hosted in their garden.

without outside help, even to cutting back perennials in fall or spreading compost during planting time. "It's the gardening that's keeps us healthy," says Anita.

As passionate about people as they are about plants, it wasn't long before Mike and Anita joined the local hosta and daylily societies. Both the Sheehans have served on the board of directors of the Western New York Hosta Society and have hosted everything from society teas to garden tours as part of the Gardens Buffalo Niagara program. They love the social aspect of these groups as much as the learning opportunities they afford and smile broadly as they talk about the many friends they have made through these organizations.

The American Hosta Society has named the hosta the "friendship plant," a title Mike and Anita clearly embrace, generously sharing their treasures as well as gratefully receiving gifts of plants from others. Indeed, that's how Anita first came to gardening, when after admiring something in a nearby garden the neighbor immediately reached for the shovel and dug out a clump to share. That's what friendship is all about, as far as this big-hearted couple is concerned. Sharing themselves, their passion, and their plants. What a wonderful way to live a life.

Playing with Color and Texture

Artistic intuition has guided the Sheehans, and they have learned, by trial and error, a few tricks to keep their ever-expanding plant collections from morphing into a mixed bag of horticultural jelly beans. Looking for specific color echoes between plants suggests a sense of belonging; combining a yellow daylily with a yellow grass, for example, can enhance both. And it's easy to create a sophisticated vignette by grouping different hostas; simply use solid colors or plants that are lightly patterned to set off more flamboyant specimens, just as you might use a wide, dark belt to separate a striped skirt from a polka dot blouse.

Flowering plants are magnificent when in bloom but can be a dreary mélange of green leaves the rest of the time. Daylilies and many flowering shrubs are examples of exactly that problem, but the Sheehans have a solution. They have underplanted their hedge of lilacs, flowering quince (*Chaenomeles speciosa*), and mock orange (*Philadelphus*), all of which look stunning in early spring, with an assortment of hostas, whose bold, brightly colored leaves add sparkle and disguise the base of these tall shrubs. In front of the hostas is a small strip of lawn before a bed of summer-blooming perennials has its moment in the spotlight. This tiered effect may be likened to a piece of framed art, the daylilies being the watercolor masterpiece, bordered by a green mat (the lawn) and accented with a colorful fillet (the hostas), before a solid frame (the shrubs) defines the limitations of the scene.

TOP LEFT Here a variegated *Hosta* 'Captain Kirk' shines a spotlight on darker notes, a color echo between the ripening berries of St. John's wort (*Hypericum*) and richly hued coral bells (*Heuchera* 'Peach Flambé').

TOP RIGHT Blue is the theme that unites these hostas so effectively. Variegated *Hosta* 'Paul's Glory' dominates the foreground, benefiting from the backdrop of a solid blue *H.* 'Big Daddy', with its tightly quilted, cupped leaves.

ABOVE A lightly ruffled reblooming daylily (*Hemerocallis* 'James Dean') looks fabulous next to golden Japanese forest grass (*Hakonechloa macra* 'Aureola'). In fall, this grass often has deep red threads running through the blades, which will draw special attention to the flower's red markings.

LEFT Summer-blooming daylilies and coneflowers benefit from a layered picture frame of green lawn, bright hosta foliage, and dense shrub border.

158   A Collector's Garden

One or two coneflowers amid a sea of daylilies is likely to be seen as an accident. By adding bold drifts of companion plants, a strategic design statement is clearly made. The Sheehans have used coneflowers and black-eyed Susans effectively for this purpose. Their depth of color matches the intensity of the daylily blooms, yet the unifying daisy shapes make them stand apart. They also have the advantage of attracting bees and butterflies by the dozen, introducing a delightful ever-changing dynamic to the scene.

The challenge when designing with a large collection of hostas is to allow each specimen to shine. Sometimes that is achieved by playing with color, but other times using contrasting foliage texture is the key. Mike has enjoyed learning more about companion plants for his many hostas, most of which feature bold texture. Finely textured ferns are a favorite, and he is especially enamored with the lattice-like detail of his Victorian lady fern (*Athyrium filix-femina* 'Cruciato-cristatum'). Both astilbe and goat's beard (*Aruncus*) have fern-like foliage; some of the dwarf forms are especially dainty, offering striking combinations with bolder companions.

Keeping Up Appearances

With three 150-foot-long borders dedicated to collections of summer perennials that will be dormant during winter, consideration needs to be given to spring, when those plants will offer little but green

TOP The texture of dwarf goat's beard (*Aruncus aethusifolius*) contrasts starkly with that of a newly planted Siberian bugloss (*Brunnera macrophylla* 'Alexander's Great'), whose silver-veined, heart-shaped leaves will eventually surpass all but the grandest of Mike's hostas in size.

ABOVE Coneflowers attract many different pollinators.

LEFT A glorious display of several different coneflowers and black-eyed Susans (*Rudbeckia fulgida* var. *sullivantii* 'Goldsturm') mingle to create substantial drifts between the daylilies; a few rogue beebalms (*Monarda*) lighten the moment.

shoots. In sunny areas, the Sheehans have planted hundreds of daffodils around the emerging daylilies, their spent foliage easily disguised as the hostas and other perennials mature: Mike's favorite hostas for sun are Sun Power (golden yellow), Liberty (green and creamy white), Sum and Substance (chartreuse), Blue Wedgewood (blue), and Krossa Regal (gray-blue). They did try tulips one year, but of the five hundred they planted only six survived. It didn't help that the tulips emerged when they were out of town, and that without regular spraying they were at the mercy of the hungry herd. "Deer love tulips," Mike says resignedly.

Shadier areas benefit from the addition of bleeding heart (*Lamprocapnos spectabilis*) and other spring ephemerals for early season interest. Gold Heart (*L. spectabilis* 'Gold Heart') is a favorite of Anita's, known for its golden leaves and pink heart-shaped flowers that dangle from the arched stems. With adequate water and afternoon shade, this foliage can persist through summer.

Finally, even with such careful attention to extending seasons of interest, a collector's garden runs the risk of being predictable. Mike and Anita have added the element of surprise while also interjecting their humor and personalities. Decorative garden art is tucked discreetly into the garden, and an assortment of birdfeeders and birdhouses keep feathered friends visiting on a regular basis.

LEFT A tin-can tin man always elicits a smile.
RIGHT Tucked into the branches of a sheltering maple, a rustic birdhouse hosts another season of inhabitants.

Top 10 Plants

RUBY GIANT PURPLE CONEFLOWER (*Echinacea purpurea* 'Ruby Giant').
This easily grown perennial is noted for its oversized daisy flowers, with ruby-pink petals and orange-brown central cone. Makes an excellent cut flower, attracts pollinators, and has a long bloom time, from summer until fall. Grows 2.5 feet tall, 1.5 feet wide. Full sun, zones 3–9. Deer resistance: B.

GUM DROP PURPLE CONEFLOWER (*Echinacea purpurea* 'Gum Drop').
Feathery pom-pom flowers are the hallmark of this cultivar. Deep pink double flowers stand tall on stout stems; at 2.5 feet tall and 1.5 feet wide, this is an ideal candidate for the middle of the border. Full sun, zones 4–9. Deer resistance: B.

VICTORIAN LADY FERN (*Athyrium filix-femina* 'Cruciato-cristatum').
The original parent fern was discovered in England in the 1860s and presented to Queen Victoria as a gift. Seedlings vary enormously in size and structure but typically are 1–3 feet tall and wide and display an intricate lattice-like detail. These ferns do best with evenly moist soil and protection from strong winds. Partial shade–full shade, zones 4–8. Deer resistance: A.

DWARF CHINESE ASTILBE (*Astilbe chinensis* var. *pumila*).
Thick stiff panicles of lavender-pink flowers appear above the mounding green foliage in summer. This spreading, herbaceous perennial, 1 foot tall and wide, thrives with even soil moisture; it has better sun tolerance than other astilbes but looks its best in partial shade. Partial shade–full shade, zones 4–8. Deer resistance: B.

PRESIDENT LINCOLN LILAC (*Syringa vulgaris* 'President Lincoln'). An outstanding variety of a beloved multi-stemmed deciduous shrub, 10 feet tall and 6 feet wide, known for its large trusses of sweetly fragrant, single Wedgwood-blue flowers that perfume the air in spring. As a hedge component, it provides seasonal screening. Full sun–partial sun, zones 3–8. Deer resistance: B.

SNEEZEWEED (*Helenium*). The identity of this variety, received as a gift, is not known, but Mardi Gras or Moerheim Beauty are similar in color and habit, the former being a little shorter. Sneezeweed is a spreading, herbaceous perennial that is easy to grow, to 5 feet tall, 3 feet wide. The colorful blooms make excellent cut flowers and attract pollinators. Full sun–partial sun, zones 3–8. Deer resistance: B.

GOLD HEART BLEEDING HEART (*Lamprocapnos spectabilis* 'Gold Heart'). With golden fern-like foliage and arching branches of dangling pink flowers, this is a colorful perennial for the shade garden. The foliage typically dies back by midsummer but may persist into fall with regular watering. Grows 2 feet tall, 3 feet wide. Partial shade–full shade, zones 3–9. Deer resistance: A.

CORAL CHARM PEONY (*Paeonia* 'Coral Charm'). Deep coral buds open into large, bowl-shaped, semi-double flowers in a softer coral-peach, fading gradually to ivory; the deep green foliage is attractive through fall. An excellent cut flower with a light fragrance. Grows 3 feet tall and wide. Full sun–partial sun, zones 4–8. Deer resistance: A.

ORCHID SATIN ROSE OF SHARON (*Hibiscus syriacus* 'Orchid Satin'). Large orchid-pink blooms, each with a deep red eye, appear in great profusion along the gracefully arching branches of this deciduous shrub, 8–12 feet tall, up to 6 feet wide. This variety does not set seed and therefore lacks the invasive tendencies of some older forms. Full sun, zones 5–9. Deer resistance: B.

DAFFODIL (*Narcissus*). Thriving in organically rich, well-drained soil, these iconic spring-blooming bulbous perennials may naturalize in the garden. Many flower types, colors, and sizes are available; plants are 0.5–2.5 feet tall and up to 1 foot wide. Full sun–partial sun, zones 4–8. Deer resistance: A.

A Hilltop Hacienda
where native and adapted plants bring Mexico to Texas

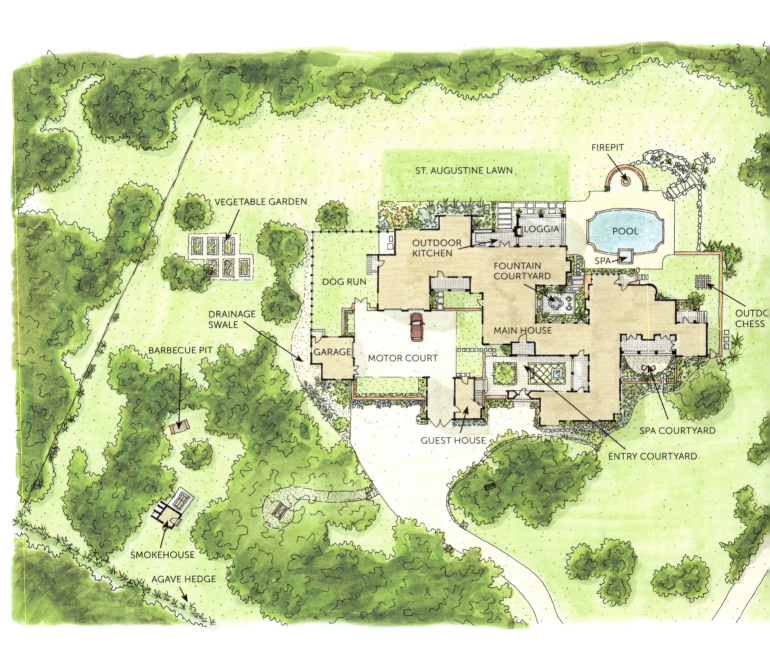

T he driveway meanders slowly upward, past windswept junipers and mounds of golden daisies that set the rocky hillside ablaze, anticipation building with every turn. A striking Spanish-style home finally comes into view, a multi-residence family compound set within a landscape that fuses the best of native Texan and adapted Mexican plants. Tall iron gates and warm-toned stone and stucco walls protect the inner dwellings and courtyards from deer, who are otherwise free to wander this 9-acre estate on the edge of Texas Hill Country.

Wide-open spaces were just one reason Richard and Margaret Lonquist fell in love with this property, a former ranch where longhorn cattle and horses used to roam. Besides boasting panoramic views, it was on a dead-end street within Austin's premier Eanes Independent School District, an important consideration for their two

QUICK FACTS
LOCATION: Austin, Texas (zone 8)
SOIL TYPE: caliche and limestone
PROPERTY SIZE: 9 acres
PROBLEM CRITTERS: deer, rabbits, squirrels, armadillos, turkeys
OTHER CHALLENGES: extreme heat and drought; flash floods; watering restrictions

DESIGN CREDITS:
home of Richard and Margaret Lonquist
designed by Tait Moring, Tait Moring & Associates

TOP Stone pillars flank the entrance to the hilltop compound. The entry court was designed to preserve as many existing live oaks (*Quercus virginiana*) as possible while allowing for adequate guest parking.

ABOVE Carefully placed lighting fixtures cast captivating shadows, creating a magical effect.

daughters. It took five years to complete the 10,000-square-foot home, designed to reflect the couple's love of both Mexico and Morocco, and it has proven since to be a great home for the family as well as for private entertaining and hosting charity functions. Tile flooring and sliding glass doors across the back of the main house seamlessly connect the indoor living spaces with an outdoor loggia, from which there is access to a large limestone patio showcasing a pool and integral spa. An adjacent semicircular stone bench wraps around a firepit, a perfect spot to toast marshmallows on a starlit evening. Options for family fun continue with an outdoor chess set just steps away under the shade of an oak tree.

Planning the Garden

The Lonquists hired landscape architect Tait Moring, who has more than thirty years' experience designing deer-resistant landscapes, during the early stages of the home's construction, and together they spent the following two years refining the plans for the garden. Both Margaret and Richard wanted to include many plants that were indigenous to Texas, especially cacti and succulents, a good choice given local watering restrictions. The only anomaly was their insistence on the inclusion of a thirsty St. Augustine lawn, which reminded them both of their childhoods in Houston. Margaret further requested that the landscape design soften the home's exterior while withstanding the regular evening visits from deer. The deer also cause considerable damage during rutting season, when they rub their antlers against the bark of young trees and even sharp-spined plants like agave. Spraying deer repellent is often recommended as a preventive measure to help plants become established, but Tait doesn't consider the ongoing maintenance of spraying worth the effort. "The deer will always win," he says.

With only an inch or two of rock-like caliche soil over solid limestone, it was necessary to bring in additional soil to create berms for planting; most of the native plants were selected for their ability to adapt to this soil once their roots became established. Erosion caused by flash flooding is a problem throughout Austin, so dry streambeds were installed to divert water away from the house and to slow its progress, allowing time for it to percolate through

RIGHT A drainage swale diverts rainwater, its edges softened with trailing lantana (*Lantana montevidensis*) and pink muhly (*Muhlenbergia capillaris* 'Regal Mist').

BELOW Giant hesperaloes (*Hesperaloe funifera*) pierce a deer-resistant groundcover of gray-woolly twintip (*Stemodia lanata*). Within the walled spa courtyard, the banana tree (*Musa*) is protected from deer.

the porous limestone before eventually draining into the Barton Creek watershed. Water runoff also posed a significant problem on the steep hillside; to reduce erosion and stabilize the slope, the landscaping crew stapled a thick, biodegradable fabric to the slope and seeded Habiturf grass over it. Habiturf is a mix of three native grasses—buffalo grass (*Bouteloua dactyloides*), blue grama (*B. gracilis*), and curly mesquite (*Hilaria belangeri*)—selected (after many years of research by the Lady Bird Johnson Wildflower Center) for its ability to thrive with minimal watering and mowing.

A lower pasture encompassing approximately three acres was briefly considered as a suitable location for a vineyard; however, when Richard realized how much work a vineyard entailed, he opted instead to sow the area with Texas bluebonnets (*Lupinus texensis*) and other native wildflowers, whose colors the Lonquists enjoy from spring through late July. Richard also loves his smokehouse, oversized though it may be, with a large pit just outside for barbecuing brisket. "It was drawn out on a cocktail napkin. Considering I have an engineering firm, it probably should have been engineered!"

The Hacienda Landscape

During construction the Lonquists took great care to protect the many live oaks on the property; these stately evergreens lend a sense of maturity to the new home and help ground it to the landscape. To add seasonal interest, the designer introduced bigtooth maple (*Acer grandidentatum*), which is native to Texas and a particular favorite of Richard's, in addition to Texas red oak (*Quercus buckleyi*) and chinquapin oak (*Q. muehlenbergii*), both of which add fall color.

With the trees in place, an assortment of native or adapted shrubs and perennials (including grasses) were layered in, offering year-round interest with spring and fall highlights. Texas mountain laurel (*Sophora secundiflora*) and yaupon holly (*Ilex vomitoria*) are two popular evergreen shrubs used for this purpose, loved for their blue spring flowers and red winter berries, respectively. To marry the Spanish

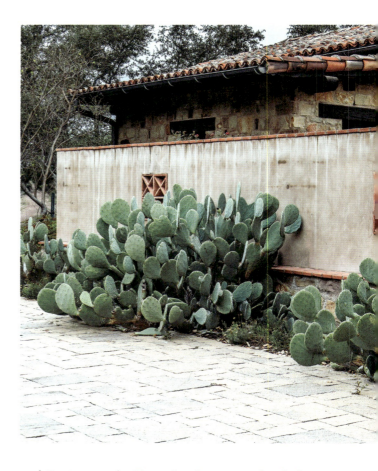

architecture to the Texas landscape, and to balance the scale of the large home, bold architectural succulents and cacti, including yuccas, agaves, and golden barrel cactus (*Echinocactus grusonii*), were incorporated throughout.

In a partially shaded bed, firecracker plant (*Russelia equisetiformis*) billows seductively at the base of three sculptural metal vessels, each planted with variegated mother-in-law's tongue (*Sansevieria trifasciata*), enjoyed as a frost-tender annual that can be overwintered indoors. The rich brown of the containers plays off the terracotta tiles that cap the sand-colored stucco walls.

The rear of the home and surrounding gardens receive unrelenting afternoon sun, placing heat tolerance high on the list of essential plant characteristics. *Muhlenbergia* 'Pink Flamingo' and other muhly grasses thrive here, and in fall, their airy seed heads soften and echo the warm stucco tones of the two-story façade.

ABOVE LEFT Near the entry gates, spineless prickly pear (*Opuntia ellisiana*) and gopher plant (*Euphorbia rigida*) combine in a monochromatic planting, the cool blue-green contrasting with the warm stucco wall.

ABOVE RIGHT Repetition of color and form, contrasting textures, and a tantalizing glimpse into an inner courtyard, where a thornless Chilean mesquite (*Prosopis chilensis*) presides, create an alluring composition.

LEFT Fire and Ice—a study in contrasting color, texture, and form using spineless prickly pear (*Opuntia ellisiana*), firebush (*Hamelia patens*), and flame acanthus (*Anisacanthus quadrifidus* var. *wrightii* 'Mexican Fire').

OPPOSITE Pink Flamingo muhly and other statuesque grasses balance the scale of an imposing façade.

ABOVE An upper outdoor pavilion offers panoramic views, carefully preserved by omitting tall plantings in the western border. A mixed bed of shrubs adds color at the base.

RIGHT Cool and hot colors collide in this combination of firebush (*Hamelia patens*) and Texas sage (*Leucophyllum frutescens*).

BELOW RIGHT A row of century plants (*Agave americana*) forms a thorny, architectural hedge.

Welcome breezes pass through an outdoor pavilion on the second floor, a perfect place for watching sunsets. A deep border at the base of the stone structure is planted with drought-tolerant shrubs, including firebush and esperanza (*Tecoma* 'Orange Jubilee'), the hot colors tempered with blue nolina (*Nolina nelsoni*) and the silver foliage of Texas sage (*Leucophyllum frutescens*), a native shrub that flourishes in hot, dry conditions and rocky soil.

Richard is especially proud of a hedge of steel-blue century plants (*Agave americana*), planted on the eastern boundary. It was inspired by a visit to the town of Tequila in Mexico, where tequila agaves (*A. tequilana*) are grown in long rows. While the deer don't eat these spiny succulents, they have damaged them during rutting season—but not so much that the Lonquists have found it necessary to use protective cages.

A Hilltop Hacienda

ABOVE LEFT Limestone-edged raised beds are individually netted to keep critters out.

ABOVE RIGHT Basil (*Ocimum basilicum*) flourishes in this climate and is seldom severely damaged by deer.

LEFT Richard constructed the hoops from plumbing parts and bird netting.

Vegetable Garden

On the southwestern side of the home, a series of six stone-framed, raised beds provide the family with a bounty of vegetables and herbs. Rather than protecting these from deer by using a tall perimeter fence, Richard, the primary vegetable gardener, uses hoops and bird netting to cover individual beds, where he grows sweet potatoes, tomatoes, jalapeno, and cabbage; Margaret likes to add zinnias and sunflowers for cutting.

While mostly successful, if the netting isn't replaced carefully, squirrels can get in and wreak havoc, together with the more athletic rabbits. At least the deer have been thwarted, discouraged perhaps by the large rosemary bushes.

Top 10 Plants

TEXAS SAGE (*Leucophyllum frutescens*). This evergreen shrub, 5–8 feet tall, 4–6 feet wide, thrives in hot, dry conditions. It has silver-gray leaves and lavender-pink flowers between summer and fall. Often called the barometer plant, as it blooms when it is about to rain and the air pressure changes. Good drainage is essential. Full sun, zones 8–10. Deer resistance: A.

FIREBUSH (*Hamelia patens*). Grown against a west-facing wall, this is an evergreen shrub in the Lonquist garden, thriving in the protected location and rocky soil, to a height of 1–3 feet and a width of 1–2 feet. Full sun–partial shade, zones 9–10. Deer resistance: A.

CENTURY PLANT (*Agave americana*). A native of Mexico, this iconic agave thrives in dry, rocky soil. The common name suggests it will live for a hundred years; in fact, once it blooms (usually in ten to twenty-five years) it dies, leaving pups around the base. The plant is 3–6 feet tall, 6–10 feet wide, and may be damaged by deer rutting against it in fall and winter. Full sun–partial sun, zones 8–10. Deer resistance: B.

PLATEAU GOLDENEYE (*Viguiera dentata*). This native annual creates a dazzling butterfly-attracting display of golden yellow flowers in fall, and the seed heads that follow the blooms are a favorite with finches. Reseeds readily; considered weedy by some. Extremely drought tolerant, thriving in rocky calcareous soils. Grows 3–6 feet tall and wide. Full sun–partial sun, zones 8–9. Deer resistance: A.

A Hilltop Hacienda

LINDHEIMER'S MUHLY (*Muhlenbergia lindheimeri*). Popular for its tall stature, fine texture, and lacy, purplish to silver-gray plumes in fall, this perennial grass is native to parts of Texas and thrives in the dry, limestone soil. Grows 3–6 feet tall and wide. Full sun, zones 7–10. Deer resistance: A.

ROSEMARY (*Rosmarinus officinalis*). A favorite for the kitchen as well as the deer-resistant landscape, this aromatic evergreen shrub thrives in hot, dry conditions. Blue spring flowers attract bees. Grows 2–6 feet tall, 2–4 feet wide. Full sun, zones 8–10. Deer resistance: A.

BIGTOOTH MAPLE (*Acer grandidentatum*). This tree's size, typically 20–30 feet tall and wide, is influenced by moisture levels, growth being reduced in drier conditions. Fall color ranges from red to yellow and orange. Trees may be tapped for their sap, which yields a syrup that equals that of the better-known sugar maple (*A. saccharum*) in quality. Full sun–partial shade, zones 5–8. Deer resistance: B.

BLACKFOOT DAISY (*Melampodium leucanthum*). Highly adaptable to arid landscapes, but the honey-scented blooms will be more prolific with regular watering. Good drainage is essential for this native Texas perennial. Grows 10–12 inches tall, 12–20 inches wide. Full sun, zones 5–10. Deer resistance: A.

TEXAS MOUNTAIN LAUREL (*Sophora secundiflora*). Violet-blue flowers, with a scent reminiscent of grape Kool-Aid, smother this multi-stemmed evergreen shrub in spring. Drought tolerant once established, this Texas native thrives in warm, arid conditions and, as a slow grower, is well suited to smaller gardens, eventually 15–20 feet tall, 10–15 feet wide. Full sun, zones 8–11. Deer resistance: A.

YAUPON HOLLY (*Ilex vomitoria*). At 10–20 feet tall and 8–12 feet wide, this evergreen shrub is popular for hedging and loved for its bright red berries, which are produced when both male and female plants are present. It adapts to both drought and poor drainage. Full sun–partial shade, zones 7–9. Deer resistance: B.

A Curated Garden
where everyday natives and choice ornamentals find common ground

Spiky, variegated agaves juxtaposed with soft blue layers of ground-hugging conifers; a topiary octopus, crafted from cotoneaster, waving its red-berried arms; the shaggy bark of silver broom (*Adenocarpus decorticans*) emerging from a dark, neatly sheared hedge of California wax myrtle (*Morella californica*)—these are just a few of the unexpected vignettes one encounters on Jim and Deborah Heg's Whidbey Island property. And a bonus: these combinations are not only artistic, they are drought- and deer-resistant. It has taken twenty-four years to get to this point, but Deborah can finally look at her garden and feel satisfied. "This is me," she says with a smile.

Even as a child Deborah knew that one day she would leave Los Angeles for somewhere "moister and greener," inspired in part by memories of family sailing trips to Whidbey Island, a popular vacation destination noted for its sandy beaches and mild maritime climate. When she

QUICK FACTS
LOCATION: Greenbank, Washington (zone 8)
SOIL TYPE: sand
PROPERTY SIZE: 21 acres
PROBLEM CRITTERS: deer, rabbits
OTHER CHALLENGES: no irrigation

DESIGN CREDITS:
home of Jim and Deborah Heg
designed by Deborah Heg

The Hegs sited their new home to take advantage of the expansive views: a place to watch the sun rise and set over the sound.

and Jim married in 1991, they began their search for "some land and a semi-rural life." They considered many places but kept returning to Whidbey, and eventually found a large parcel of land in Greenbank, a small community of approximately 250 residents in the center of the island. Jim, who grew up in a tiny farm town in Ohio, asked, "Do you have any idea how much twenty-one acres *is*?" Yet he was the one who declared they'd found the spot.

They were drawn to the southwest-facing views over Puget Sound, looking toward Marrowstone Island and the Olympic Mountains, in addition to the quiet, park-like forest setting. The property had been logged in the 1920s, so sufficient time had passed for stands of native Douglas fir (*Pseudotsuga menziesii*), western red cedar (*Thuja plicata*), and bigleaf maple (*Acer macrophyllum*) to regenerate. The discovery of three old-growth trees dating back some 350 years

was an exciting bonus. A driveway, installed in the 1940s, led through the property from the main road down to the beach, and vestiges of the old logging road promised to make access for construction vehicles easier.

That was the good news. Unfortunately, the only buildings that existed were a small summer cabin (where the totem pole now presides) and an equally tiny guest house, neither of which epitomized the home this couple envisioned.

Jim and Deborah worked with general contractor Bob Arndt to design a Craftsman-style house sensitive to its surroundings. The resulting home appears to nestle into the land with its long, low horizontal lines and deep eaves. The Hegs selected a stain for the cedar siding to mimic the color of the surrounding tree trunks, and the dark green roof blends with the forest canopy.

ABOVE The low-profile Craftsman home appears to emerge from the garden, the forest-inspired colors of the siding and roof blending sympathetically with the environment.

RIGHT A weathered totem pole, carved by local artist Glen Russell, is framed by the feathery plumes of toe-toe grass (*Austroderia richardii*) and the coppery seed heads of shrubby hare's-ear (*Bupleurum fruticosum*).

BELOW A bigleaf maple (*Acer macrophyllum*), protected during construction, stands sentry at a fork in the drive.

Such construction does not happen overnight, however, so after some initial clearing of the building site, Deborah began researching how best to minimize trauma to those trees selected to remain, knowing that compaction by materials and vehicles would pose a significant threat to their long-term survival. Their solution was to leave buffer trees in place, specimens that were slated for eventual removal but could remain until construction was complete, creating a physical barrier between the vehicles and the trees they wished to protect.

Today a new, fern-lined drive sweeps past majestic conifers, mature bigleaf maples, and ancient stumps, connecting a large three-car garage adapted as Jim's office and man cave, continuing past a guest house and an art studio, before reaching the main house.

Master Planning

Creating a landscape plan for such a large property can be intimidating, but Deborah possessed a clear vision right from the beginning. She had been inspired by the park-like settings of Bloedel Reserve (Bainbridge Island, WA) and Stanley Park (Vancouver, BC) in addition to the work of Capability Brown. The dual sense of openness and enclosure, created by the juxtaposition of rolling lawns and forested areas, was something she knew she wanted to emulate. Placement of the main lawn was dictated by necessity: a septic drainfield together with a reserve area was needed adjacent to the main house and studio. With that in place, Deborah turned her attention to the forest. She noted that several trees grew in discrete groups or islands; once defined by clearing out the underbrush, these became distinctive spaces

Garden visitors are enticed to explore the forest trails by passing through a rusted metal pergola, jokingly nicknamed the bus stop. It is flanked by a Japanese maple and a weeping Persian ironwood (*Parrotia persica* 'Pendula') planted almost twenty years ago.

ABOVE Bronze rabbit sculptures by Georgia Gerber cause fewer problems than their living counterparts.

LEFT Not afraid to mix patterns, Deborah combines a variegated silverberry (*Elaeagnus pungens* 'Hosoba-fukurin') and white-variegated sedge (*Carex*), the contrast in texture ensuring success.

BELOW A manzanita (*Arctostaphylos densiflora* 'Howard McMinn'), carefully limbed up to reveal its smooth red bark, is now a sculptural focal point on the terrace through which water views can be glimpsed.

A Curated Garden

separated by areas of open grass, such that one moved comfortably through a series of experiences, from the wilder, outer reaches of the forest through dappled glens toward the main house.

A 22-foot-wide reflecting pool, installed even before the house was complete, is the focal point of the largest clearing. The inspiration for it came from a photograph in *The Farmhouse* by Chippy Irvine (Bantam, 1987), which depicted a low-profile reflecting pool set in a forest clearing; a second book, *The Bloedel Reserve* by Lawrence Kreisman (Bloedel Reserve, 1988), introduced the notion of surrounding a pool with a sheared hedge, to combat the feeling of its "leaking space." Deborah was struck by that concept and decided to imitate it in her own garden, using a low stone wall as the pool's boundary line rather than a hedge.

Behind the home, generously proportioned granite terraces take advantage of the water view;

Morning light hits the still surface of the reflecting pool, a moment of pure magic.

creeping thyme (*Thymus serpyllum* 'Elfin') softens the flagstone pavers underfoot. Low stone walls define the space here, too, and provide a spot to display several bronze sculptures by local artist Georgia Gerber.

Deborah and Jim were excited to see deer on the property and never considered installing a fence to exclude them. "I wanted to be all hippy-dippy about it," she laughs. "You know, one with the flora and the fauna—we would all live together, and the deer would like me!" She also naively assumed that if she selected native plants for her landscape, the deer would leave them alone, before realizing that many of those plants were both attracting and sustaining the deer.

Observation was key in those early days. Deborah noted the deer's preferred path through the property and that everything they tried to grow in that area would get eaten, while the same plants a few yards away remained untouched. In fact, the resident herd were especially fickle in their tastes, feasting on native flowering currant (*Ribes sanguineum*), salal (*Gaultheria shallon*), and wintergreen (*G. procumbens*) in some areas but not others, decimating serviceberry (*Amelanchier*) and mock orange (*Philadelphus*), and even eating the foliage of madrones (*Arbutus menziesii*) in hard winters.

Damage caused by the deer was twofold; taste-testing in spring and summer, and abrasion during the fall rutting season. To address the former, Deborah sprays Bobbex on the plants closest to the house; she hopes to teach the deer, when they are young, that something doesn't taste good, so they won't try it again. Alas, Bobbex "stinks to high heaven," so she adds a few drops of rosemary and camphor essential oils to the mix; both are known to have deer repellent properties and are marginally helpful in disguising the commercial spray's sulfurous smell.

Farther away from the house, Deborah did not feel the time, energy, or cost involved in spraying was justified and preferred to grow only reliably deer-resistant trees and shrubs in those areas, ignoring any minor damage. Jim, however, thought otherwise. He loves snowball viburnums (*Viburnum opulus*) and wanted some by his office. The deer were delighted at this tasty addition to their diet, and so began a battle of the wills that endured for several years as Jim tried to ward them off with repellent sprays. Eventually, he admitted defeat and removed the remaining, severely disfigured shrubs.

To protect against antler damage, Deborah places wire frames around low-growing junipers and other vulnerable plants. The metal cages are easy to erect and remarkably discreet, visually disappearing at a distance. In spring they can be removed and stored, although she presses a few into service to protect emerging hostas. "I was told the deer don't eat the tougher, larger-leaved hostas," she says. "That's nonsense!" Without protection the juicy new shoots are just too tempting for the herd.

In order to grow fruit, vegetables, and cutting flowers (including dahlias, a personal favorite of Deborah's), the Hegs incorporated a large (90 feet by 120 feet) vegetable garden into the master plan. This

LEFT Sturdy yet discreet wire frames are pushed into the ground and used to protect vulnerable plants. The deer tried rutting against the silverberry (*Elaeagnus ×ebbingei* 'Gilt Edge') only once, but Deborah isn't taking any chances.

BELOW The only permanent barrier Deborah and Jim use to exclude deer is around their vegetable garden. The double gates are wide enough to allow a tractor to enter, but the dense 8- to 9-foot-high hedge thwarts even the most athletic of deer.

A broad swath of lawn leads from the terrace to a quiet seating area. A deep, dense border of juniper (*Juniperus chinensis* 'Shimpaku', *J. sabina* 'Tamarix'), dwarf Norway spruce (*Picea abies* 'Gregoryana Parsonii'), and New Zealand holly (*Olearia macrodonta*), among others, creates a mosaic of color and texture without obstructing the view.

veritable fortress has sturdy cedar gates worthy of a castle, while the 8- to 9-foot-tall hedge of Portuguese laurel (*Prunus lusitanica*) not only excludes the deer but make a strong design statement, its calming, horizontal line a counterpoint to the vertical tree trunks of the surrounding forest. Within this enclave Deborah can grow a wide selection of produce and armfuls of blooms for the home, safe from the inquisitive deer.

The Hegs have relaxed into their complementary roles maintaining the garden: Deborah is the primary designer and gardener, Jim the big-picture landscape manager, overseeing the lawns and taking pride in keeping the drives clear of debris. With the benefit of time and experience, their plant knowledge has increased, enhanced by friendships with Holly Turner, co-creator of Froggwell Gardens on Whidbey Island, and other gardening luminaries. Proximity to Heronswood (Kingston, WA), a botanical garden and nursery curated by plantsman Dan Hinkley, also fueled Deborah's passion for the rare and unusual.

She loves nothing more than coming home with a box full of treasures and wandering around the garden trying to decide where best to place them. However, she admits that were she to start over, the plant selection would be much simpler, her self-confessed plant lust having added to the already somewhat high-maintenance design. "In this climate, the exuberance of growth is constant," she acknowledges. Weeding is less of a problem since little bare earth is exposed, but the shaping and trimming of trees and shrubs is very time consuming. Yet it is Deborah's artistic pruning and limbing up of select specimens that has created intriguing portals, exposed interesting bark, or revealed unusual forms. A retired interior designer, she has translated her love of color and textured fabrics into an intricately woven tapestry of plants, each one thoughtfully placed to ensure a pleasing result.

The gardens are designed primarily for spring and fall interest, with summer and winter being quieter. Deborah loves every shade of green, especially in spring, appreciating it as a foil for the flowers of her favorite hellebores (*Helleborus*), barrenworts (*Epimedium*), crocus, hardy cyclamen, and other ephemerals. In fall the evolution through a kaleidoscope of fiery colors is always breathtaking. A vine maple (*Acer circinatum*) in the front garden blazes when backlit by the autumnal sunshine, bathing the home's interior in a warm glow. Another favorite is a scarlet oak (*Quercus coccinea*), whose foliage turns bright red. Notably absent from her combinations are white flowers, although she made an exception for a large stand of matilija poppies (*Romneya coulteri*), which remind her of her California childhood, and black mondo grass (*Ophiopogon planiscapus* 'Nigrescens'), a very popular evergreen presence in Pacific Northwest gardens. Deborah explains her edict against black mondo: "It's just too dark. Here the light is very harsh, almost glaring. With such high contrast, dark colors are lost in the shadows."

Naturalistic and predominantly native plantings at the outer reaches of the property transition gradually into more contemporary mixed borders adjacent to the house, where architectural agaves thrive in the heat and well-drained soil that has been amended with pumice, protected from heavy winter rains by the extended eaves. Deborah's quest for unusual specimens has resulted in striking vignettes that are an especial source of inspiration and encouragement to other deer-challenged homeowners. Far from giving up on her garden dreams, Deborah seems instead to have had her creativity fueled by her cloven-hooved challengers.

Together Jim and Deborah have created a garden, not for entertaining large groups, but for personal sanctuary. It isn't a place cloaked in silence, however. Eagles cry out as they soar overhead, black-capped chickadees splash noisily in basalt birdbaths, and far below gentle waves glissando across the beach. Certainly, in hindsight, there are things the Hegs would have done differently, from planting shade trees adjacent to the patio, to siting the vegetable garden much closer to the house. But overall, after more than two decades of planning, planting, nurturing, and editing, Deborah is satisfied with the way most of her plant combinations have matured and coalesced. The garden now provides the serenity she sought, and she relishes the days when she can be completely immersed in it.

Front Garden

Initially, Deborah enlisted the help of a professional to design the beds closest to the home, requesting a focus on native plants; as a recent California transplant, she was unfamiliar with the local plant palette and felt that natives would be well adapted to the maritime climate and sandy soil of Whidbey Island. But while the designer's inclusion of vine maples and scarlet oak resonated with the Hegs, the underplanting of rhododendrons and a groundcover of kinnikinnick (*Arctostaphylos uva-ursi*) did not; rather than blending into the surroundings, the somewhat stark design appeared incongruous with the larger, more naturalistic landscape. Additionally, the stone walls were initially kept free of any encumbering foliage, whereas Deborah felt they needed to be softened.

Deborah began devouring gardening books, and as her knowledge of plants and understanding of garden design grew, she was able to modify the original plan. Specifically, she began to research broadleaf evergreens that might do well in the front beds. Seeking plants for winter interest and fragrance, Deborah incorporated Oregon grape (*Mahonia*), sweetbox (*Sarcococca*), and drooping fetterbush (*Leucothoe fontanesiana*), all of which are generally deer-resistant and thrive in partial shade. She also discovered the world of conifers, adding Siberian cypress (*Microbiota decussata*) and other unusual specimens, often mail-ordered from specialist nurseries, as well as a few, more pedestrian conifers: she has no snobbery when it comes to junipers!

When a home has such strong horizontal lines, choosing the scale of adjacent plantings can be tricky. Deborah needed larger, bold foliage, such as comfrey (*Symphytum*), mayapple (*Podophyllum*), and even grapevines (*Vitis*), to stand out against the distinctive architecture, realizing that "just jamming things in" can impair the visual balance.

ABOVE The bold texture and distinctive variegated foliage of a comfrey (*Symphytum* ×*uplandicum* 'Axminster Gold') contrasts pleasingly with the fine-textured carpet of insideout flower (*Vancouveria hexandra*), a Pacific Northwest native.

RIGHT Close to the front door a planting of glossy, dark green sweetbox (*Sarcococca ruscifolia*) greets winter guests with sweet perfume, as well as providing a backdrop for the more colorful variegated knotweed (*Persicaria virginiana* 'Painter's Palette') and spreading English yew (*Taxus baccata* 'Repandens').

BELOW Deborah has created a dramatic framework of evergreen foliage using conifers, broadleaf evergreens, evergreen perennials, and succulents. A prostrate deodar cedar (*Cedrus deodara* 'Feelin' Blue') and blue nest spruce (*Picea mariana* 'Ericoides') frame a variegated century plant (*Agave americana* 'Opal') growing in a pot. The yellow-needled foliage of a Japanese cedar (*Cryptomeria japonica* 'Sekkan-sugi') stands out against the dark cedar shingles; layers of broadleaf shrubs, including Winter's bark (*Drimys winteri* var. *chiloense* 'Pewter Pillar') and alpine mint bush (*Prostanthera cuneata*) complete the scene.

For years Deborah tried to grow pigsqueak (*Bergenia cordifolia* 'Tubby Andrews') in the front garden, the bold yellow and green variegated foliage an eye-catching counterpoint to the dark siding of the home. The resident deer, however, saw this as an invitation to their personal snack bar. Eventually, Deborah gave up the fight and dug them out. In their place, she has added wall germander (*Teucrium chamaedrys*) as an interrupted, low-growing hedge, whose ease of shearing and reliable deer resistance make it a much lower-maintenance option.

Deborah discovered that a calm palette and repetition of color and shape helped to unify the complex mosaic of plants in the front garden. There are no jarring high-contrast color combinations, even in fall. Rather, the design is held together by subtle shades of green and blue, interspersed with burgundy or copper accents. Likewise, rather than confusing the eye, the many varieties of finely textured grasses and grass-like foliage in this relatively small space read as a deliberate repetition of a soft, mounding form.

To the right of the front door, a border extends to wrap around the western end of the house. Here Deborah indulges her love of ferns, which thrive in the dappled shade cast by the mature maples and scarlet oak, combining them with hostas, Japanese shrub mint (*Leucosceptrum japonicum* 'Golden Angel'), and other perennials. Standing guard over the scene is a whimsical green cat made by Deborah. "Only a mother could love it," she laughs, hence its relegation to the shadows.

A small dish rock—and the cat—are nestled within a bed of mixed ferns, including soft shield fern (*Polystichum setiferum*) and royal fern (*Osmunda regalis*).

Top 10 Plants

NEW ZEALAND HOLLY (*Olearia macrodonta*). A holly-like broadleaf evergreen with gray-green foliage and clusters of fragrant, white daisy-like flowers in summer. Tolerates hard pruning but can be allowed to grow as a small tree, 6–10 feet tall, 5–6 feet wide, exposing the peeling bark. A robust shrub for coastal locations and deer-prone gardens. Full sun, zones 8–11. Deer resistance: A.

MOUNTAIN HOLLY (*Olearia ilicifolia*). Not as tolerant of heavy soils as others in the genus but still an evergreen shrub worthy of garden space. It differs from other species by having highly serrated gray-green leaves and distinctive black stems but shares the profusion of white daisy-type flowers in late spring. Grows 4–6 feet tall, 6 feet wide. Full sun, zones 8–10. Deer resistance: A.

SHRUBBY HARE'S-EAR (*Bupleurum fruticosum*). A tough evergreen shrub, 4 feet tall and wide, that tolerates salt spray and is adaptable to most soil types and conditions (except boggy) and either drought or irrigation. The dill-like flowers and subsequent coppery seed heads are popular with flower arrangers, while the prominently veined, glossy green leaves are a pleasing foil for more colorful companions in the landscape. Full sun, zones 6–10. Deer resistance: A.

DAUB'S FROSTED JUNIPER (*Juniperus ×pfitzeriana* 'Daub's Frosted'). A spreading, low-growing conifer, 1–2 feet tall and 3–6 feet wide, noted for the sunny yellow tips on its pendulous blue-green branches, presenting an overall frosted appearance. Drought tolerant once established. Full sun, zones 4–9. Deer resistance: B.

ACE BARBERRY (*Berberis wilsoniae* 'Ace'). Steel-blue foliage with dusky plum overtones makes this a colorful shrub for the mixed border. Yellow summer flowers are followed by rosy pink fruit, borne on thorny branches that the deer typically ignore. Drought tolerant once established. Grows 3–5 feet tall, 3–4 feet wide. Full sun–partial shade, zones 5–9. Deer resistance: A.

SILVER BROOM (*Adenocarpus decorticans*). A profusion of golden yellow blooms adorns the branches of this deciduous shrub in late spring and early summer, almost masking the finely textured green foliage. When limbed up as a small tree, the shaggy, peeling bark is revealed, introducing shades of soft butterscotch and creamy white to the scene. Needs excellent drainage to thrive. Grows 6–9 feet tall and wide. Full sun–partial sun, zones 8–10. Deer resistance: A.

GILT EDGE SILVERBERRY (*Elaeagnus ×ebbingei* 'Gilt Edge'). A large evergreen shrub, 8–10 feet tall and wide, that catches the eye no matter where it is planted. Striking green and gold variegated leaves, overlaid with silver scales, somewhat mask the tiny white flowers in fall. The flowers may not be memorable, but their fragrance most certainly is. Adaptable to drought and many soil types. Full sun–partial shade, zones 7–9. Deer resistance: A.

SPARKLE BARBERRY (*Berberis thunbergii* 'Sparkle').
In spring and summer this deciduous shrub forms a deep green mound of thorny foliage, 3–4 feet tall and wide, studded with yellow flowers. It is fall when it truly lives up to its name, the leaves a fiery mix of red, orange, and gold, enhancing the jewel-like red berries. Drought tolerant once established. Full sun–partial sun, zones 4–8. Deer resistance: A.

JELENA WITCH HAZEL (*Hamamelis ×intermedia* 'Jelena').
This deciduous, vase-shaped shrub, 8–12 feet tall and wide, is a highlight of the fall and winter garden. Fall color runs the gamut from purple to crimson and gold, each leaf showing a slight variation in pattern, while spidery copper-colored winter blooms perfume the air at a time of year when much of the garden is quiet. Full sun–partial shade, zones 5–8. Deer resistance: B.

EVERGREEN HUCKLEBERRY (*Vaccinium ovatum*).
A Pacific Northwest native that prefers moisture-retentive soil but adapts to drought once established and tolerates clay soil, providing drainage is adequate. The small edible berries appear in late summer. Grows 6 feet tall and wide. Adapts to full sun or woodland conditions, even tolerating full shade, although it will not flower or fruit as well in that location. Zones 7–9. Deer resistance: A.

A Suburban Retreat
bringing Colorado to New Jersey

Scott Bradley knew exactly what he wanted: a secluded retreat in the Colorado mountains. The problem? His home and third-generation family business was in suburban New Jersey. After studying psychology in Colorado, which is where he and Lezli met, Scott headed back east. "She's a city girl, and I'm a country boy at heart," says Scott, "but this home has been a good compromise." Lezli, a Colorado native, worked in the fashion industry in New York before opening her own women's clothing boutiques in New Jersey twenty-five years ago; the interior of their home reflects her artistic sensibility.

The couple moved to their center-hall Colonial in Chatham when their son, Duncan, was in first grade. It was a perfect family home for entertaining, especially after Lezli opened up the cramped floor plan to improve flow. A two-story addition created a new master suite and

> **QUICK FACTS**
> LOCATION: Chatham, New Jersey (zone 6)
> SOIL TYPE: loam
> PROPERTY SIZE: 0.56 acre
> PROBLEM CRITTERS: deer, rabbits, chipmunks
> OTHER CHALLENGES: steep hillside; shade from large trees; poor drainage

DESIGN CREDITS:
home of Scott and Lezli Bradley
designed by Susan Cohan, Susan Cohan Gardens

rooftop deck, which is now a favorite oasis; they enjoy the treetop view while savoring a glass of wine as the sun goes down. "It has become a little house that lives large," says Lezli.

It soon became apparent, however, that the landscape was in need of attention. The crudely built patio and a 6-foot-tall retaining wall constructed from railroad ties were far from inspiring, while on the hillside below, towering 100-foot tulip trees (*Liriodendron tulipifera*) and a mélange of weeds jostled for dominance. The entire back garden was dark, forbidding, and unusable. Oddly enough, despite the many trees, the home felt completely exposed to the surrounding neighbors, since the canopy was so high.

ABOVE RIGHT The traditional architecture of the Bradleys' home belies the updated, contemporary interior, which Lezli designed.

OPPOSITE ABOVE An intimate shaded patio leads off the guest suite on the lower level.

OPPOSITE BELOW Spanning the length of the house, a generous dining patio is accessible from the back door and family room.

RIGHT Tiered stone walls and a series of wide steps tamed the steep hillside. The boxwood, rhododendrons, and other large shrubs were transplanted from other parts of the property.

BELOW The moss rock used for the steps is encrusted with lichen.

A Suburban Retreat

ABOVE Layered plantings, including several eastern redbuds (*Cercis canadensis* 'Forest Pansy'), now offer only intriguing glimpses of the house.

OPPOSITE Opting for simplicity and harmony with the adjacent fern garden, Lezli planted the tall charcoal vessels on the dining patio with shade-loving Boston ferns (*Nephrolepis exaltata* 'Bostoniensis').

The Illusion of Seclusion

The couple owned a vacation home in the Colorado mountains for eighteen years, and that, together with a sentimental attachment to the place where they first met, drove Scott's desire to recreate that idyll in suburban New Jersey. The first step was to replace the retaining wall with natural stone, and for this the Bradleys turned to landscape contractor Frank Scheppe, who also crafted two new patios and replaced the original wooden steps with wide, shallow slabs of moss rock, making access to the lower garden much safer.

When it came time to address the privacy concerns and fine-tune Scott's ideas, however, Frank introduced them to landscape designer Susan Cohan. To resolve the fishbowl effect, Susan considered the view from every window of the house and planted trees to block those sight lines. The result is a remarkable sense of seclusion despite the close proximity of their neighbors, enhancing the illusion of being in a private mountain retreat.

Scott stressed the need for a low-maintenance planting plan, as both he and Lezli work seven days a week; he also requested native or naturalized plants whenever possible ("We can't do annuals") and the inclusion of a dry pond. This feature, as well as reminding them of their Colorado hiking adventures, would serve the practical purpose of mitigating water runoff, which causes significant erosion of the hillside during heavy storms. Lezli, meanwhile, decided that the secret to a successful marriage was for her to do all the interiors and leave the outside to Scott. The patios were their shared middle ground.

Wildlife Challenges

Chatham may be only a short commuting distance from metropolitan New York, but it is also an area where black bears, wild turkeys, foxes, rabbits, chipmunks, and—especially on this property—deer roam freely and often. Anything included in the garden must be able to withstand frequent browsing by deer; even plants that are considered deer-resistant elsewhere get decimated here, which accounts for the early failure of oakleaf hydrangeas (*Hydrangea quercifolia*) and variegated Japanese aucuba (*Aucuba japonica* 'Gold Dust'), both of which Susan was sure would be ignored. Yet she remains philosophical: "The presence of deer is just an opportunity to do something else. The best design is about solving problems in practical ways."

Her perseverance and experience paid off. The majority of the plants have proven reliably deer-resistant, although the hostas and rhododendrons are sprayed with deer repellent, a task which takes ten minutes once a month. Scott has learned to warn his neighbors who are downwind of the property in advance and apologizes profusely for the unpleasant smell, which can linger for up to three hours.

Colorado Touchstones

At the Bradleys' request, the designer incorporated a Colorado spruce (*Picea pungens* 'Fat Albert') and an abundance of Colorado blue columbines (*Aquilegia coerulea*), the Colorado state flower. Their strategic placement adjacent to the dry pond mimics their native habitat among craggy boulders, where the profusion of their flowers is a highlight of alpine springs. But trying to train the landscaping maintenance crew to recognize the delicate columbine seedlings was an exercise in futility. Scott even tried flagging the seedlings to highlight their presence, but too often they fell victim to string trimmers. Now he selectively hand-weeds that area himself.

These initial plant choices inspired Susan's use of purple and yellow, two complementary colors that always look good together. She included lots of yellow flowers, including daffodils, Japanese kerria

An existing rock-filled depression was expanded and enhanced with larger boulders and a granite bridge to create a naturalistic dry pond and creek, designed to capture excess rainwater. A small stone bench nestles within the waves of golden Japanese forest grass (*Hakonechloa macra* 'Aureola') and Colorado blue columbines (*Aquilegia coerulea*), now setting seed for next year's display.

(*Kerria japonica*), and yellow barrenwort (*Epimedium ×versicolor* 'Sulphureum'), with punctuations of Siberian iris (*Iris* 'Caesar's Brother') and assorted purple foliage. Yet the palette is not strictly limited to these two colors: per Scott's request, the design included white azaleas (*Rhododendron* 'Delaware Valley White'), whose fuzzy leaves likely contribute to their being deer-resistant, and the pink *Astilbe* 'Rheinland', which is particularly good as a cut flower.

Fall is a highlight in the garden. The vanilla fragrance of bugbane (*Actaea simplex* Atropurpurea Group 'Hillside Black Beauty') fills the crisp autumn air, the tall spires of white flowers standing high above the dark foliage, and as the trees change color and leaves begin to fall, there is a sudden flare of rich cobalt-blue from several stands of autumn monkshood (*Aconitum carmichaelii* 'Arendsii'). The color is electrifying and totally unexpected at that time of year.

TOP LEFT A waterfall of golden Japanese forest grass cascades onto the steps that link the lower lawn to the dry pond, an inspired juxtaposition of soft and hard.

TOP RIGHT A hint of purple: the heart-shaped leaves of Forest Pansy eastern redbud (*Cercis canadensis* 'Forest Pansy'), Scott's favorite tree, seen from below.

ABOVE Rheinland astilbe is known for its long-lasting summer display of pink blooms, while the plant's upright form and fresh green foliage add structure to the garden from late spring until fall.

When he saw a large stand of ferns in the woods, Scott was inspired to recreate that look in his garden. Now he has a veritable hillside of them. By transplanting vigorous ostrich ferns (*Matteuccia struthiopteris*) from other parts of the property onto what he previously considered "an ugly hill," he has transformed the unusable space into a feathery carpet over which an inviting hammock is slung. Ferns also play a role in the front garden, breaking up a large swath of pachysandra (*Pachysandra terminalis*) near the driveway. "We don't *love* the pachysandra, but it works," says Scott, noting that it has taken a while to become established since the tulip trees soak up so much water.

A layered planting approach softens the western borders as well. Beneath the tulip trees, shade-loving hellebores (*Helleborus* 'Ivory Prince') and coral bells add color and bold texture to a groundcover of sweet woodruff (*Galium odoratum*), which is slowly becoming established. Susan has been surprised to observe that deer do not seem to take an interest in dark-leaved coral bells (*Heuchera villosa* 'Palace Purple', *H.* 'Obsidian').

LEFT In the fern garden, a hammock offers a secluded spot for reading. The ostrich ferns (*Matteuccia struthiopteris*) die to the ground in winter and don't need cutting back or raking out, their biomass naturally revitalizing the soil.

BELOW Scott has achieved the Colorado forest feeling he sought with myriad shades of green. Clockwise from left: Japanese maple (*Acer palmatum* 'Viridis'), ostrich fern (*Matteuccia struthiopteris*), variegated Solomon's seal (*Polygonatum odoratum* var. *pluriflorum* 'Variegatum'), and Virginia sweetspire (*Itea virginica* 'Little Henry').

A Suburban Retreat

TOP Coppery fronds of autumn ferns (*Dryopteris erythrosora*) and red-stemmed northern lady ferns (*Athyrium filix-femina* var. *angustum* 'Lady in Red') punctuate the tight rosettes of pachysandra (*Pachysandra terminalis*), an evergreen and reliably deer-resistant groundcover.

ABOVE Evergreen coral bells (*Heuchera*), hellebores (*Helleborus*), and sweet woodruff (*Galium odoratum*) are slowly carpeting the ground beneath trees on the western perimeter.

Looking to the Future

Scott realizes the native dogwood trees (*Cornus florida*) are beginning to fail and so has introduced some saplings to take their place, eventually, continuing the cycle of life and preserving the tiered woodland effect they love so much. Whenever possible, he tries to help the trees recover if they get damaged, carefully pruning away injured limbs to improve airflow and light. Tropical storms, hurricanes, and recent heavy snowfalls have given him plenty to do; severe winds sheared off the uppermost treetops and ripped limbs from redbuds and dogwoods, leaving behind a motley collection of Dr. Seuss–like casualties. These wounded warriors have a story to tell, Scott feels, one he hopes is of adjustment and endurance; they are rather like humans in that regard, as they struggle to adapt to adversity in order to survive.

Nurtured over the many years, the naturalistic landscape now stands out for its lushness and richly textured design. Every day as he's driving home, Scott slows down, appreciating the wooded view of the dry pond, the deck, the beautiful stonework. Colorado has indeed melded into this New Jersey hillside, and that fusion is picture-perfect.

Top 10 Plants

GOAT'S BEARD (*Aruncus dioicus*). Fluffy, cream-white panicles of flowers appear in early summer on a plant 4–6 feet tall and 2–4 feet wide. Partial shade is usually recommended, but with persistent moisture, this perennial can withstand full sun. Zones 4–8. Deer resistance: C.

AUTUMN MONKSHOOD (*Aconitum carmichaelii* 'Arendsii'). This perennial blooms later than other species; the deep blue flowers are reminiscent of hoods worn by medieval monks and make an eye-catching display in early fall. Does best in moisture-retentive soils, to 2–4 feet tall, 1–1.5 feet wide. Full sun–partial shade, zones 3–7. *Warning*: all parts of this plant are highly toxic; wear gloves when handling and do not grow adjacent to edible plants or where children play. Deer resistance: A.

LITTLE HENRY VIRGINIA SWEETSPIRE (*Itea virginica* 'Little Henry'). This deciduous shrub offers attractive foliage, racemes of fragrant white flowers in spring, and fiery fall color. Deer may nibble the flowers. Grows 1.5–2 feet tall, 2–2.5 feet wide, forming dense colonies by suckering. Full sun–partial shade, zones 5–9. Deer resistance: B.

FOREST PANSY EASTERN REDBUD (*Cercis canadensis* 'Forest Pansy'). Heart-shaped purple leaves that glow when backlit are the hallmark of this deciduous tree; fall color is typically gold with orange and reddish purple highlights. Grows 20–30 feet tall, 25 feet wide. Plant in well-drained soil. Full sun–partial shade, zones 5–9. Deer resistance: C.

OSTRICH FERN (*Matteuccia struthiopteris*). A clump-forming, upright, deciduous fern with a shuttlecock-type appearance, each long frond reminiscent of a feathery ostrich plume. Grows 2–3 feet tall and wide in cultivation; in the wild, in ideal conditions (rich, moist, shaded), these form dense colonies, 3–6 feet tall, 5–8 feet wide. Partial shade–full shade, zones 3–7. Deer resistance: A.

LADY IN RED NORTHERN LADY FERN (*Athyrium filix-femina* var. *angustum* 'Lady in Red'). Tolerant of heavy shade, rabbits, and deer, this deciduous fern also does surprisingly well in drier soils. Lacy green fronds carried on a distinctive red central stalk. Grows 1.5–2.5 feet tall and wide. Partial shade–full shade, zones 4–8. Deer resistance: A.

VARIEGATED SOLOMON'S SEAL (*Polygonatum odoratum* var. *pluriflorum* 'Variegatum'). A spreading, shade-loving perennial, 2–3 feet tall, 1 foot wide, with arching stems of oval, green leaves edged in white. In spring, pairs of white bell-shaped flowers hang beneath the foliage. Fall color is a soft, buttery yellow. Although cited as being occasionally severely damaged by deer, this has not been the Bradleys' experience. Partial shade–full shade, zones 3–8. Deer resistance: C.

MOSS GREEN PRIVET HONEYSUCKLE (*Lonicera pileata* 'Moss Green'). A low-growing evergreen or semi-evergreen shrub with glossy, bright green leaves on a dense twiggy structure, 1–2 feet tall and 5–8 feet wide. Drought tolerant in shade. Full sun–partial shade, zones 6–8. Deer resistance: B.

SWEET WOODRUFF (*Galium odoratum*). Fragrant, white spring flowers are attractive, and the foliage is aromatic when dried. Grows 0.5–1 foot tall. Well behaved in drier soils but can become an invasive thug in moist, shady areas, where it will spread by creeping roots as well as by self-seeding. Partial shade–full shade, zones 4–8. Deer resistance: A.

JAPANESE KERRIA (*Kerria japonica*). This lax, multi-caned shrub can be trained to scramble against a support or allowed to form a large, informal mound, 5–10 feet tall, 6–10 feet wide. Single yellow flowers appear in late spring; the bright green leaves turn yellow in fall. Partial shade, zones 4–9. Deer resistance: B.

Deer-Resistant Container Gardens

Entertaining friends on the patio, surrounded by exuberantly colorful pots, needn't be forfeited just because deer may crash the party. Use these creative container designs to explore the many possibilities, or use them as a springboard for your own ideas.

Some plants included here may tempt an especially inquisitive deer. In such cases, homeowners often resort to an occasional spritz with a proprietary deer repellent as an insurance policy. You may prefer to substitute another plant, and, where possible, suggestions have been given.

Tutti-Frutti Fiesta

PETRA CROTON (*Codiaeum variegatum* 'Petra'). Evergreen foliage boldly variegated in fiery shades of red, orange, and gold is the hallmark of this exotic houseplant, which thrives outdoors in warmer climates. Where hardy, it can grow 7–10 feet tall and 4–6 feet wide. This needs moist but well-drained soil with protection from afternoon sun to prevent scorching and maintain the vibrant fiesta colors. Zones 9–11. Deer resistance: A.

DIANA BROMELIAD (*Guzmania* 'Diana'). Native to tropical forests, bromeliads make exciting additions to shaded summer containers in more temperate climates. The potting soil should be able to drain freely but remain slightly moist to the touch. Diana has luminous yellow bracts that rise above the strap-like green basal foliage. Grows 12–20 inches tall, 12–15 inches wide. Zones 9–12. Deer resistance: A.

SWITCH BROMELIAD (*Guzmania* 'Switch'). Known for its dark red flowering bracts, Switch grows 12–20 inches tall and 12–15 inches wide. Zones 9–12. Deer resistance: A.

CLASSIC BROMELIAD (*Guzmania* 'Classic'). Orange flowering bracts are the calling card of this cultivar, which grows 12–20 inches tall and 12–15 inches wide. Zones 9–12. Deer resistance: A.

GOLDEN CREEPING JENNY (*Lysimachia nummularia* 'Aurea'). Easily grown in moist, shaded conditions, this chartreuse perennial is popular in containers for its trailing habit and in the landscape as an evergreen groundcover. In late spring it bears buttercup-yellow flowers. Grows 6 inches tall and spreads 1.5 feet or more. Zones 3–8. Deer resistance: B.

Let the party begin! Vibrant tropical colors, reminiscent of juicy, fruit-flavored candies, are showcased to perfection in the deep plum container. Setting the planter on an orange pedestal allows the golden creeping Jenny to trail with abandon. Site: partial shade.

Stylish Duet

PANDOLA FLAMINGO FLOWER (*Anthurium* 'Pandola'). Grown for its glossy pink flower (technically a spathe, a type of bract), this popular houseplant is a bold addition to shade gardens and containers during summer months. Grows 1.5 feet tall and wide. Zones 10–11. Deer resistance: B.

GHOST FERN (*Athyrium* 'Ghost'). An upright herbaceous fern that has silver fronds with a burgundy midrib. Grows 2 feet tall and wide in full shade or partial shade. Zones 4–8. Deer resistance: A.

HIMALAYAN MAIDENHAIR FERN (*Adiantum venustum*). Lacy layers of bright green held on wiry black stems. This evergreen fern spreads slowly to form a feathery groundcover. Grows 6 inches tall and 3 feet wide in partial shade. Zones 5–8. Deer resistance: A.

BOSSA NOVA PURE WHITE BEGONIA (*Begonia boliviensis* Bossa Nova 'Pure White'). A continuous summer display of pendulous white blooms on coral-pink stems earns this drought-tolerant tuberous begonia a place in shady containers and baskets. Grows 16 inches tall, 16–20 inches wide. Zones 7–10. Deer resistance: B.

SPIDER'S WEB JAPANESE ARALIA (*Fatsia japonica* 'Spider's Web'). This evergreen shrub thrives in the shade, its youngest foliage exhibiting the most dramatic speckled coloration. Grows 3–5 feet tall and wide. Zones 7–11. Deer resistance: B.

SILVER SHADOW ASTELIA (*Astelia* 'Silver Shadow'). A drought-tolerant evergreen perennial with bold, metallic silver leaves. Grows 2–3 feet tall and 2–4 feet wide. Zones 7–11. Deer resistance: B.

LAGUNA WHITE LOBELIA (*Lobelia erinus* 'Laguna White'). A low-maintenance summer annual with small white flowers throughout the season. Grows 8–12 inches tall and trails or spreads up to 2 feet. Zones 9–11. Deer resistance: A.

Shimmering shades of silver, white, and celadon form a delicate framework accented by flamingo-pink. From the dramatic foliage of the speckled Japanese aralia to the broad metallic blades of the astelia and froth of white lobelia blooms, this potted duet will certainly garner the attention of your guests—but thankfully not the deer. Site: partial shade.

Whales on Wheels

VANZIE WHALE'S TONGUE AGAVE

(*Agave ovatifolia* 'Vanzie'). Eventually reaching 3–4 feet tall and wide, this handsome agave is more adaptable to damp cool climates than other blue agaves and is drought tolerant once established. Protection from antler damage in fall and winter may be necessary. Zones 7–9. Deer resistance: B.

A rusted tractor rim takes on a new life as a shallow container for a large whale's tongue agave. The low, wide profile of this repurposed vessel is the perfect complement to the broad succulent, a striking specimen with its deeply channeled ice-blue leaves. Site: full sun.

Banana and Blueberry Sundae

RED ABYSSINIAN BANANA (*Ensete ventricosum* 'Maurelii').
An outstanding ornamental banana noted for its burgundy-flushed leaves and dark midribs. Protect from afternoon sun and strong winds. Grows 10 feet tall and wide, shorter if grown as an annual. Zones 9–11. Deer resistance: B.

PURPLE FOUNTAIN GRASS (*Pennisetum setaceum* 'Rubrum').
Foxtail-like plumes in shades of pink, burgundy, and tan rise above the loose fountain of burgundy foliage in summer. May be invasive in some areas. Grows 3–4 feet tall and wide. Zones 8–11. Deer resistance: A.

MEXICAN PETUNIA (*Ruellia simplex*).
Commonly grown as a bog or water plant, it is also adaptable to average garden conditions and containers. A fast-growing perennial, this blooms best in full sun although it will also grow in partial sun. Invasive in some areas; check with your local extension office before planting. Zones 8–10. Deer resistance: B.

PETUNIA (*Petunia*).
The variety featured in this design is unknown, but any white mounding or trailing petunia, mini-petunia, or million bells (*Calibrachoa*) could be used to the same effect. Grows 1 foot tall and wide. Although not a favorite of deer, the petunias may tempt them, so spraying with a deer repellent would be wise, or switch in something more reliably deer-resistant, such as a white lobelia. Annual. Deer resistance: C.

A tall red-leaved banana takes center stage in this blueberry-blue pot, its rich colors repeated by the grasses and Mexican petunias that fill out the middle of the design. Add a scoop of simple white petunias, and this delightful yet understated summer dessert is complete. Site: partial sun.

Controlled Burn

ORANGE ROCKET BARBERRY (*Berberis thunbergii* 'Orange Rocket'). Noted for its upright growth, the ruby-red foliage turns orange in fall. Before planting, check to see if barberries are invasive in your area. Grows 4 feet tall, 2–4 feet wide. Zones 4–9. Deer resistance: A.

BEYOND BLUE BLUE FESCUE (*Festuca glauca* 'Beyond Blue'). A versatile evergreen grass that retains its powder-blue color better than the species. Grows 9 inches tall, twice as wide. Drought and heat tolerant. Zones 4–9. Deer resistance: A.

GOLDEN OREGANO (*Origanum vulgare* 'Aureum'). A robust, golden groundcover that is both ornamental and edible. Grows 1–3 feet tall and 1 foot wide. Zones 5–9. Deer resistance: A.

CAMPFIRE FIREBURST BIDENS (*Bidens* 'Campfire Fireburst'). Heat tolerant and vigorous, the flame-orange blooms appear in waves from spring until fall. Grows 1–1.5 feet tall, 1.5–2 feet wide. Zones 9–11 or enjoy as an annual. Deer resistance: A.

SILVER FALLS DICHONDRA (*Dichondra argentea* 'Silver Falls'). Metallic silver foliage that looks fabulous cascading from hanging baskets and containers. May also be used as a drought-tolerant groundcover. Zones 10–11 or enjoy as an annual. Deer resistance: A.

Sizzling shades of red, orange, and gold are tempered by silvery blues in this reliably deer-resistant and sun-tolerant design. Several bidens accentuate the foliage framework with sparks of flame-orange flowers, adding a note of informality to the display and softening the more structured placement of their companions. Site: full sun.

Fusion Design

CHERRIES JUBILEE FALSE INDIGO (*Baptisia* 'Cherries Jubilee').
The maroon and gold bicolor blooms of this low-maintenance, drought-tolerant perennial attract bees and butterflies before maturing to form decorative seedpods. Grows 3 feet tall and wide. Zones 4–9. Deer resistance: A.

ASCOT RAINBOW SPURGE (*Euphorbia* ×*martinii* 'Ascot Rainbow').
Attractive gold and green variegated foliage blushes pink at the tips, while the flowering stems are held high. Drought tolerant once established. Grows 18–20 inches tall and wide. Zones 5–9. Deer resistance: A.

TEQUILA SUNRISE MIRROR PLANT (*Coprosma* 'Tequila Sunrise').
An evergreen shrub in mild climates, the glossy foliage emerges emerald-green and gold, marbled with sunset shades that intensify in cold weather. Where winter hardy, this forms an upright pyramid, 5 feet tall and wide. Zones 9–10. Deer resistance: A.

MEXICAN FEATHER GRASS (*Nassella tenuissima*).
A finely textured ornamental grass that sways in the gentlest breeze. Grows 2 feet tall and wide. Before planting, check to see if this is invasive in your area. Zones 6–10. Deer resistance: A.

DIAMOND FROST SPURGE (*Euphorbia* 'Diamond Frost').
A heat- and drought-tolerant annual that blooms in a confetti-like haze all summer. Grows 1.5 feet tall and wide. Zones 10–11 or enjoy as an annual. Deer resistance: A.

PURPLE QUEEN (*Tradescantia pallida*).
A vigorous, spreading groundcover with deep purple, succulent foliage; may also bear small dark pink flowers. Drought and heat tolerant. Reaches 1–1.5 feet tall and will spread to 3 feet or more, but easily trimmed in a container. Zones 7–11. Deer resistance: B.

LEMON THYME (*Thymus citriodorus*).
A semi-evergreen groundcover, this culinary herb is loved for its small green and yellow variegated leaves, which exude a strong lemon fragrance when crushed. Grows up to 1 foot tall and spreads 1.5 feet or more. Zones 5–8. Deer resistance: A.

ROYALE ROMANCE VERBENA (*Verbena* 'Royale Romance').
This vigorous, drought-tolerant annual fills the middle of the design and drapes casually over the sides of the container, weaving through its companion plants with ease. The rich burgundy flowers have a tiny white eye. Grows 0.5–1 foot tall, 1.5–2 feet wide. Zones 8–11. Deer resistance: A.

ANGEL WINGS SENECIO (*Senecio candicans* 'Angel Wings').
Felted, silver, heart-shaped leaves with delicately scalloped edges make an exquisite display in containers, where this foliage plant will grow 10–12 inches tall and wide. Drought and salt tolerant. Zones 7–10. Deer resistance: A.

There's something for everyone in this design: wispy grasses and prairie perennials for the meadow lover, flowers for the traditionalist, fragrance for the romantic, and stunning foliage for those who prefer a more contemporary flair. Since most of the drought-tolerant plants can be reworked as landscape perennials or houseplants when summer is over, this blended design is also a wise investment. Site: full sun.

Deer-Resistant Container Gardens 213

Wild Tresses

FOXTAIL FERN (*Asparagus densiflorus* 'Myersii'). Long, plume-like stems give this fern a sculptural form. Drought tolerant once established, this easy-care fern is popular as a houseplant in colder areas. The species self-seeds and can be mildly invasive in subtropical and tropical climates, but this selection produces fewer berries so is less of a problem. Grows 1–3 feet tall, 3–4 feet wide. Zones 9–11. Deer resistance: A.

FRAZZLE DAZZLE DYCKIA (*Dyckia choristaminea* 'Frazzle Dazzle'). Reminiscent of an unkempt black mondo grass, this mini bromeliad thrives in well-drained soils and light shade to full sun. The fragrant yellow blooms that appear in April on 8-inch stems are a bonus. Grows in a clump 4–6 inches tall, 10–15 inches wide. Zones 8–10. Deer resistance: A.

Soft, ringlet-like plumes cascade around a voluptuous charcoal-gray pot: a foxtail fern turned seductress. To one side, a shorter pot gives serious attitude, the unkempt, grass-like foliage of a black dyckia suggesting neither brush nor hair gel will tame the defiant spikes. With these two pot personalities at the front door, you know the homeowner must have a great sense of humor. Site: partial shade.

Summer Blues

CLARITY BLUE FLAX LILY (*Dianella* 'Clarity Blue'). An exciting new hybrid that appears to be more tolerant of cold temperatures and wet soils than other cultivars. Upright, broad blue blades are evergreen in frost-free climates. Drought tolerant once established, this will grow 2–2.5 feet tall and 1.5–2 feet wide. Zones 8–11. Deer resistance: A.

MEERLO LAVENDER (*Lavandula allardii* 'Meerlo'). The fragrant variegated foliage is reason enough to grow this heat- and humidity-tolerant lavender; the pale blue flowers are a welcome bonus. Grows 2–3 feet tall, 2–2.5 feet wide. Zones 9–10. Deer resistance: A.

AMETHYST FALLS OREGANO (*Origanum* 'Amethyst Falls'). A charming ornamental oregano with cascading hop-like bracts in lavender and chartreuse. Drought tolerant once established. Grows 15 inches tall and 1.5 feet wide. Zones 5–9. Deer resistance: A.

Blue-toned foliage is the common thread in this waterwise design, enlivened by the broad creamy variegation of Meerlo lavender. Completing the scene are the hop-like lavender flowers of the Amethyst Falls oregano, which tumble casually over the container rim. Site: full sun.

Sleeping Tiger

TIGER EYES SUMAC (*Rhus typhina* 'Tiger Eyes'). Feathered foliage opens chartreuse, draping languidly from fuzzy pink stems; in fall it matures to gold with orange-red highlights and sets the world on fire. May sucker so best confined to a container. Grows 6 feet tall and wide. Deer resistance appears to vary with herd. Zones 4–8. Deer resistance: A–C.

FRAGRANT BLUE HELIOTROPE (*Heliotropium arborescens* 'Fragrant Blue'). Intensely fragrant, deep purple blooms appear throughout summer. Grows 1–1.5 feet tall, 10–14 inches wide. Zones 9–11 or enjoy as an annual. Deer resistance: A.

QUICKSILVER HEBE (*Hebe pimeleoides* 'Quicksilver'). Tiny blue-gray leaves along wiry black stems give this evergreen shrub an airy, layered look. Grows 1 foot tall and 2–3 feet wide. Zones 8–11. Deer resistance: B.

BONFIRE BEGONIA (*Begonia boliviensis* 'Bonfire'). Long, fluted orange flowers cascade from this sun-tolerant tuberous begonia from late spring until fall. Grows 2–3 feet tall and wide. Zones 9–11 or enjoy as an annual. Deer resistance: B.

SUNDEW SPRINGS CREEPING JENNY (*Lysimachia* 'Sundew Springs'). A popular trailing annual, with golden foliage and buttercup-yellow blooms from late spring until fall. Grows 10–12 inches tall, 12–20 inches wide. Zones 9–11 or enjoy as an annual. Deer resistance: B.

PURPLE HEART PURPLE QUEEN (*Tradescantia pallida* 'Purple Heart'). Grown primarily for its purple foliage, this tender perennial also blooms, with small lavender flowers. Grows 1–1.5 feet tall, 16 inches wide. Zones 8–11. Deer resistance: B.

SCARLET LOTUS VINE (*Lotus berthelotti* 'Scarlet'). Typically grown as a trailing annual for its feathery blue-green foliage and the unusual flame-like red flowers. Grows 6–10 inches tall, 2 feet wide. Zones 9–11 or enjoy as an annual. Deer resistance: A.

The pendulous orange blooms of the tuberous begonia can be relied upon for months of color, as can the grape-hued heliotrope flowers, which add fragrance to the summer show. But this brooding display truly roars into life in fall, when the Tiger Eyes sumac awakens and burns bright in shades of red, orange, and gold. Site: full sun.

Tropical Romance

GIANT TARO (*Alocasia macrorrhizos*). In its natural habitat, this may grow 12–15 feet tall and 6–10 feet wide but is likely to be half that when grown as an annual. Needs regular water and fertilizer to thrive, and does best when protected from direct sun and strong winds. Zones 9–11. Deer resistance: B.

DIAMOND FROST SPURGE (*Euphorbia* 'Diamond Frost'). This heat- and drought-tolerant plant is usually grown as a summer annual, relied upon to produce clouds of white blooms throughout the season. Grows 1.5 feet tall and wide. Zones 10–11. Deer resistance: A.

VARIEGATED HEBE (*Hebe* 'Variegata'). Creamy white and green leaves and compact habit make this evergreen shrub a popular choice for both containers and the landscape. Purple flowers in summer attract bees and butterflies. Zones 8–10. Deer resistance: B.

MARGARITA SWEET POTATO VINE (*Ipomoea batatas* 'Margarita'). A vigorous trailing plant with heart-shaped chartreuse leaves. May scorch in full sun but generally considered suitable for partial shade–partial sun. Deer may eat these if a repellent spray is not used—replace with deadnettle (*Lamium maculatum*) or Siberian bugloss (*Brunnera macrophylla*), both of which are shade tolerant and more reliably deer-resistant, if preferred. To 10 inches tall and 6 feet long. Annual. Deer resistance: D.

Add a twist to the tropical flavor of giant taro by underplanting it with a soft froth of Diamond Frost spurge, whose abundant, tiny white blooms lend a distinctive feminine touch. Variegated hebe keeps up the monochromatic scheme with crisp green and white foliage, while chartreuse sweet potato vine can be relied upon to cascade over the edge of the pot. An occasional spritz of deer repellent on the sweet potato vine may be wise, as some herds find it tasty. Site: partial shade.

Winter Fragrance

EVERGREEN HUCKLEBERRY (*Vaccinium ovatum*). A versatile evergreen shrub, native to the Pacific Northwest. Its upright habit makes it suitable for containers or the landscape, where it will adapt to full sun or deep shade. White spring flowers are followed by edible black berries, and the new foliage emerges bronze, maturing to dark green. Grows 6 feet tall and wide in full sun, twice that in full shade, but it can be sheared for size or shape. Zones 7–9. Deer resistance: A.

VARIEGATED WINTER DAPHNE (*Daphne odora* 'Aureomarginata'). The perfume from a single pink-flushed white flower can scent an entire room, while the gold-edged evergreen leaves provide year-round interest in the landscape or container. Grows 3–4 feet tall and wide in partial sun–partial shade. Zones 7–9. Deer resistance: A.

HGC CHAMPION HELLEBORE (*Helleborus* ×*ericsmithii* 'HGC Champion'). Leathery evergreen foliage is a foil for the large creamy white flowers, flushed deep pink on the reverse, which appear in late winter. Grows 8–12 inches tall, 1.5 feet wide, in full or partial shade. Zones 5–8. Deer resistance: A.

MOSSY SOFT SHIELD FERN (*Polystichum setiferum* 'Plumosum Densum'). A delightful evergreen fern with remarkable symmetry and texture. Grows 20 inches tall, 1 foot wide, in full or partial shade and moisture-retentive soil. Zones 5–9. Deer resistance: A.

BLACK MONDO GRASS (*Ophiopogon planiscapus* 'Nigrescens'). Strappy near-black leaves form an evergreen carpet in sun or shade, although protection from hot afternoon sun is advised. Lavender summer blooms are followed by black berries on this clump-forming grass, which grows 5–6 inches tall and wide. Zones 6–10. Deer resistance: A.

Evergreen, colorful, fragrant, and deer-resistant—who said you can't have beautiful winter pots by your front door? In spring it would be wise to dismantle the design and transplant the individual plants into the landscape, where they can mature to their full size, but be sure to site the fragrant daphne where you can enjoy its intoxicating spicy perfume on a cold winter's day. Site: partial shade.

Mixed Spice

MADEIRA CANNA LILY (*Canna* 'Madeira'). A dwarf canna sporting fresh green leaves and flaming red blooms edged with gold. Grows up to 3 feet tall and 1 foot wide. Zones 8–11. Deer resistance: B.

AFRICAN SUNSET CANNA LILY (*Canna* 'African Sunset'). Boldly variegated leaves in shades of orange, maroon, green, and gold are the colorful foil for the large orange flowers. Grows 4–6 feet tall, 3 feet wide. Zones 8–11. Deer resistance: B.

CELEBRATION BLANKETFLOWER (*Gaillardia* ×*grandiflora* 'Celebration'). Rich red flowers appear throughout summer on this compact herbaceous perennial, whose deer resistance appears to vary considerably with the herd. Deadheading encourages rebloom. Drought tolerant once established. Grows 14–16 inches tall and wide. Zones 5–9. Deer resistance: A–C.

PINK CARMINE CROCOSMIA (*Crocosmia* 'Pink Carmine'). Salmon-pink blooms appear from midsummer through fall on this herbaceous perennial. Grows 2 feet tall and 10 inches wide. Zones 6–11. Deer resistance: B.

LEMONY LACE ELDERBERRY (*Sambucus racemosa* 'Lemony Lace'). Forming a feathery yellow mound, this scorch-resistant elderberry will shine in the garden. White spring blooms, a red edge to newly emerging foliage, and red autumnal berries add to the display. Benefits from pruning after bloom to improve shape. Grows 3–6 feet tall and about as wide. Zones 3–7. Deer resistance: A.

INFERNO MIRROR PLANT (*Coprosma* 'Inferno'). Where hardy, this glossy evergreen shrub will grow 4–5 feet tall and wide. Even as an annual, it's worth seeking out for its red, orange, and green variegated foliage, though the shrub will be much smaller. Zones 9–11. Deer resistance: A.

BANDANA ROSE LANTANA (*Lantana camara* 'Bandana Rose'). Bright pink and sunny yellow flowers attract hummingbirds and butterflies on this

Sandwiched between the tropical canna leaves and cut-leaf golden elderberry is a spicy floral feast of blanketflower, lantana, sage, and crocosmia. Pulling all these sassy blooms together in a single shiny leaf is the aptly named Inferno mirror plant. The feathery lotus vine and dark leaves of Persian Chocolate creeping Jenny add the final flavors as they spill out of the pot. Site: full sun.

variety, which is almost sterile. Grows 1–2 feet tall and 2–3 feet wide as an annual, larger where it is hardy. Zones 9–11. Deer resistance: A.

SCARLET LOTUS VINE (*Lotus berthelotti* 'Scarlet'). Vivid scarlet blooms and feathery blue-green foliage combine delightfully to trail from hanging baskets and containers. Grows 6–10 inches tall and spreads 2 feet. Zones 9–11 but usually enjoyed as an annual. Deer resistance: A.

PERSIAN CHOCOLATE CREEPING JENNY (*Lysimachia congestiflora* 'Persian Chocolate'). Dark chocolate leaves and golden blooms make this trailing plant a popular addition to containers, but it may also be grown as a groundcover in the landscape, to 6 inches tall and 1–3 feet wide in moisture-retentive soil. Zones 6–9. Deer resistance: A.

Vintage Charm

SUMMER RUFFLE ROSE OF SHARON (*Hibiscus syriacus* 'Summer Ruffle'). Attractive variegated leaves set this compact hibiscus apart from the crowd. Ruffled, lavender-pink blooms charm in late summer. Grows 3–4 feet tall and wide. Zones 5–9. Deer resistance: B.

SEÑORITA ROSALITA SPIDER FLOWER (*Cleome* 'Señorita Rosalita'). A huge improvement over older varieties, this sterile, non-sticky, non-thorny, non-stinky hybrid is a mass of pink blooms all summer on a compact plant, 2–4 feet tall and wide. Annual. Deer resistance: A.

LO & BEHOLD PINK MICRO CHIP BUTTERFLY BUSH (*Buddleja* Lo & Behold 'Pink Micro Chip'). A noninvasive mini version of the ever-popular butterfly bush with lightly fragrant, orchid-pink blooms that attract bees, butterflies, and hummingbirds. Grows just 2 feet tall and wide. Zones 5–9. Deer resistance: A.

PLATINUM BEAUTY LOMANDRA (*Lomandra longifolia* 'Platinum Beauty'). Delicately variegated, grass-like foliage looks stunning in the landscape or container. Drought tolerant once established. Grows 2–3 feet tall and wide. Zones 8–10. Deer resistance: A.

CHEF'S CHOICE ROSEMARY (*Rosmarinus officinalis* 'Chef's Choice'). With its compact mounding habit, this intensely aromatic and tasty herb is equally at home in the landscape, container, or edible garden. Blue flowers attract pollinators in spring. Grows 1–1.5 feet tall and 1 foot wide. Zones 7–11. Deer resistance: A.

PURPLE THREADLEAF JOSEPH'S COAT (*Alternanthera ficoidea* 'Purple Threadleaf'). A popular summer annual loved for its narrow, burgundy foliage. Tolerant of full sun in all but the hottest climates, as well as partial shade. Grows 8–12 inches tall and wide. Deer resistance: B.

HAWAII BLUE FLOSS FLOWER (*Ageratum houstonianum* 'Hawaii Blue'). Periwinkle-blue flowers on a compact summer annual, 6 inches tall and wide. Deer resistance: B.

Blend ornamental shrubs, colorful annuals, and edible herbs in a vintage-style container for a planter that will delight hummingbirds, butterflies, bees, and the chef in your family—but will be ignored by inquisitive deer. This charming combination of foliage and flowers would be a welcome addition to any summer patio. Site: full sun.

MAGILLA BEEFSTEAK PLANT (*Perilla* 'Magilla'). Reminiscent of a sun-tolerant coleus, this summer annual is popular for its bold, colorful foliage. Grows 2–3 feet tall, 1–1.5 feet wide. Deer resistance: B.

PINK STRIPE PURPLE QUEEN (*Tradescantia pallida* 'Pink Stripe'). Irregularly striped, succulent foliage and small lavender-pink flowers make this a popular, frost-tender groundcover where it is hardy, or houseplant and summer annual in colder climates. Grows 1–1.5 feet tall, 1.5–2 feet wide. Zones 7–10. Deer resistance: B.

LEMON THYME (*Thymus citriodorus*). Tiny green and yellow variegated leaves have a strong lemon fragrance when crushed. Evergreen in mild winters, this culinary herb makes an attractive groundcover. Lavender blooms appear in summer, attracting bees. Grows up to 1 foot tall, spreads 1.5 feet or more. Zones 5–8. Deer resistance: A.

Sparks May Fly

FIRECRACKER PLANT (*Russelia equisetiformis*). This perennial is evergreen in mild climates, and its cascading habit is ideal for containers. Tubular flowers, which appear spring through fall, are a favorite with hummingbirds. Best suited to drier climates, where its exuberant growth, to 2–4 feet tall and wide, is slowed. May be invasive in hot, humid climates. Zones 9–11, but root hardy in zone 8, where it quickly rebounds after a light frost. Deer resistance: A.

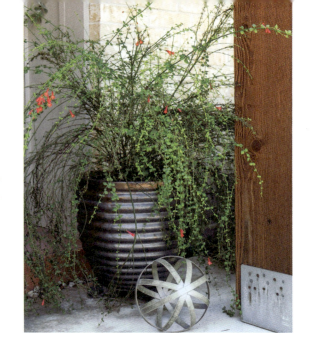

Exploding from the contemporary gray-glazed container in a shower of coral sparks, the soft, arching branches of a firecracker plant add a sense of vibrancy to this porch setting. The tubular blooms appear like tiny flames along the almost leafless stems, playing off the color of the adjacent brick and copper underglaze of the pot. Site: full sun–partial shade.

Country Re-Mix

LIL' MISS SUNSHINE BLUEBEARD (*Caryopteris ×clandonensis* 'Lil' Miss Sunshine'). Aromatic golden foliage on a compact deciduous shrub is reason enough to grow this drought-tolerant beauty—the masses of pollinator-attracting blue flowers in late summer seal the deal. Grows 2–2.5 feet tall and wide. Zones 5–9. Deer resistance: A.

PEARL GLAM BEAUTYBERRY (*Callicarpa* 'Pearl Glam'). White flowers in late summer are followed by metallic purple berries on this compact, dark-leaved shrub. Grows 4–5 feet tall and wide. Zones 5–8. Deer resistance: B

HAPPY FACE YELLOW BUSH CINQUEFOIL (*Potentilla fruticosa* 'Happy Face Yellow'). Fresh green foliage sets off the abundant yellow flowers that appear continuously from late spring until the end of summer. Grows 2–3 feet tall and wide. Low maintenance and drought tolerant. Zones 2–7. Deer resistance: A.

DRAGON'S BREATH CELOSIA (*Celosia argentea* var. *cristata* 'Dragon's Breath'). Fiery magenta-red plumes and scorched burgundy foliage make a

The country garden gets a makeover in this lively combination, where foliage is just as important as flowers. Mixing colorful shrubs with long-blooming perennials and annuals ensures a garden-like appearance that is easy to care for. The compact golden bluebeard may hold center stage, but the deep purple foliage of the beautyberry promises multi-season interest, with summer flowers and fall berries. Also offering darker shades are the plumes and foliage of the mini Chinese fringe flower and the striking celosia. Providing contrast and acting as a skirt to the display is a variegated California lilac. The compact bush cinquefoil can be relied upon to produce a profusion of yellow blooms throughout the summer. Site: full sun.

statement. A sun-loving annual that grows 2 feet tall, 16 inches wide. Deer resistance: B

JAZZ HANDS MINI CHINESE FRINGE FLOWER

(*Loropetalum chinense* 'Jazz Hands Mini'). Burgundy foliage and deep pink blooms. At just 10–12 inches tall and up to 3 feet wide, this dwarf cultivar is perfect for containers. Zones 7–9. Deer resistance: B.

POWWOW WILD BERRY PURPLE CONEFLOWER

(*Echinacea purpurea* 'PowWow Wild Berry'). Tolerant of low water and poor soils, this compact, easy-care perennial with bold pink daisy flowers will grow 16–20 inches tall and 16 inches wide. Zones 3–9. Deer resistance: B.

LOLLIPOP VERBENA

(*Verbena bonariensis* 'Lollipop'). A dwarf version, 2–3 feet tall and wide, of the popular tall verbena, perfect for containers or the front of a border. Stout stems support flat heads of lavender blooms that attract bees, butterflies, and hummingbirds. Reliably drought tolerant. Zones 6–11. Deer resistance: A.

HIGHLIGHTS CALIFORNIA LILAC

(*Ceanothus griseus* var. *horizontalis* 'Highlights'). Glossy, two-tone evergreen foliage with pale blue spring flowers. Suitable for use as a drought-tolerant groundcover. Grows 2–3 feet high, 5–6 feet wide. Zones 8–10. Deer resistance: B.

Snake Charmer

MOTHER-IN-LAW'S TONGUE

(*Sansevieria trifasciata*). An easy-to-grow houseplant that does well outdoors in summer, this popular variegated foliage plant prefers dry soil and a shaded location. Grows 2–4 feet tall and half as wide. Zones 10–12. Deer resistance: A.

Rising from a weathered metal vessel like a quiver of venomous cobras, the stiff, succulent leaves strike a bold architectural statement against the pale stucco wall. Horizontal dark green bands give the impression of snakeskin on this distinctive foliage plant, whose main charm lies in its ability to withstand less than hospitable conditions. Site: partial shade.

Child's Play

ORANGE ROCKET BARBERRY (*Berberis thunbergii* 'Orange Rocket'). Foliage opens vivid red in spring and matures to deep wine before turning orange in fall. This upright barberry grows 4 feet tall, 2–4 feet wide. Zones 4–9. May be invasive in some areas. Deer resistance: A.

MERCURY RISING TICKSEED (*Coreopsis* 'Mercury Rising'). Tolerant of deer, heat, humidity, and some drought, this low-maintenance perennial features deep red daisy flowers with a yellow eye throughout the summer. Blooms best in full sun. Grows 1–1.5 feet tall, twice as wide. Zones 5–9. Deer resistance: A.

SUNSHINE PRIVET (*Ligustrum sinense* 'Sunshine'). A versatile shrub that brings golden yellow evergreen foliage to the garden and can be used as a low hedge, container element, or landscape specimen. In winter it often takes on an orange hue. Grows 3–6 feet tall, 3–4 feet wide. Zones 6–10. Deer resistance: B.

BLACK TOWER ELDERBERRY (*Sambucus nigra* 'Black Tower'). This dark-leaved columnar selection adapts well to container culture, offering flat sprays of pink flowers in spring. Thrives in moist or average soils; may spread by suckers or self-seeding. Grows 6–8 feet tall, 3–4 feet wide. Zones 5–8. Deer resistance: A.

SUNBINI CREEPING ZINNIA (*Sanvitalia procumbens* 'Sunbini'). A popular annual, this heat-tolerant groundcover produces an abundance of yellow daisies throughout the summer. Grows 6–10 inches tall and spreads up to 20 inches wide. Zones 9–11. Deer resistance: B.

AMAZON SUNSET LOTUS VINE (*Lotus* 'Amazon Sunset'). Feathery silver-green foliage and vivid red flowers make this cascading summer annual a perennial favorite for baskets and containers. Grows 8–10 inches tall and trails up to 2 feet. Zones 9–11. Deer resistance: A.

JAZZ HANDS MINI CHINESE FRINGE FLOWER (*Loropetalum chinense* 'Jazz Hands Mini'). An exciting dwarf selection, 10–12 inches tall and 3 feet wide, with deep burgundy foliage and vibrant pink blooms. Zones 7–9. Deer resistance: B.

Kindergarten memories of fat brushes and generous pots of paint come to mind, as we were encouraged to swirl primary colors together and observe what happened. Color schemes don't have to be complicated to be effective as this design shows, with flowers and foliage in shades of red and yellow blending in a bold orange pot, surrounded by Blue Star junipers (*Juniperus squamata* 'Blue Star'). Site: full sun–partial sun.

Making an Entrance

FINE LINE BUCKTHORN (*Rhamnus frangula* 'Fine Line'). The feathery green foliage on this upright deciduous shrub turns yellow in fall. Grows 5–7 feet tall and 2–3 feet wide. Zones 2–7. Deer resistance: A.

PURPLE PREFERENCE SPURGE (*Euphorbia* 'Purple Preference'). Wine-red stems, smoky purple foliage, and lime-green flowers in spring are the key features of this shrubby cultivar, which is also mildew resistant. Grows 2–3 feet tall and wide in full or partial sun and well-drained soil. Zones 7–9. Deer resistance: A.

TECHNO BLUE LOBELIA (*Lobelia erinus* 'Techno Blue'). This summer annual produces an abundance of cobalt-blue flowers and, unlike many lobelias, thrives in heat. Grows 0.5–1 foot tall, 1–3 feet wide. Deer resistance: A.

LIMONCELLO BARBERRY (*Berberis thunbergii* 'Limoncello'). A compact variety whose foliage emerges vivid golden yellow with unusual red margins, turning to fiery shades of pink, red, orange, and green in fall. Grows 3–4 feet tall and wide. Zones 4–7. Deer resistance: A.

MIGHTY VELVET LAMB'S EARS (*Stachys* 'Mighty Velvet'). Velvety, silver foliage grows quickly into large clumps up to 1.5 feet tall and wide. Prefers full sun or partial shade and low water. Zones 7–9. Deer resistance: A.

PERSIAN CHOCOLATE CREEPING JENNY (*Lysimachia congestiflora* 'Persian Chocolate'). Dark leaves and buttercup-yellow blooms make this semi-evergreen groundcover a winner. Thrives in moist shade or full sun. Grows 6 inches tall and spreads 1–3 feet. Zones 6–9. Deer resistance: A.

BRIGHT LIGHTS YELLOW AFRICAN DAISY (*Osteospermum* 'Bright Lights Yellow'). This cheerful bloomer offers an abundance of yellow daisies on a mounding plant in full or partial sun. Enjoy as an annual in cooler climates, where it will grow 8–12 inches tall and wide. Zones 9–11. Deer resistance: A.

SILVER FALLS DICHONDRA (*Dichondra argentea* 'Silver Falls'). A popular annual, this silver-leaved trailing plant will also root into the ground to form a shimmering, heat- and drought-tolerant groundcover. Zones 10–11. Deer resistance: A.

VARIEGATED POTATO VINE (*Solanum jasminoides* 'Variegata'). The yellow and green foliage of this semi-evergreen twining vine brightens sunny baskets, containers, and trellises. White flowers appear in summer. In ideal conditions and with support, it can climb 15–20 feet high. Zones 9–11. Deer resistance: A.

Entry designs need to be tall enough to garner attention as guests walk by, yet avoid obstructing access to the door. The slender silhouette of this buckthorn is ideal for such a setting, and its spotted bark ensures winter interest. Combined here with an evergreen spurge, a yellow-variegated barberry, and the felted silver leaves of lamb's ears, it needed only a few extra flowering annuals to spill over the edges of the large container for summer interest. Site: full sun.

Patchwork Pot

ANGYO STAR FATSHEDERA

(×*Fatshedera lizei* 'Angyo Star'). A cross between English ivy and Japanese aralia, offering the best of both, with large, glossy, variegated foliage and an upright habit that can be trained or supported to cover a screen or allowed to scramble along the ground, as preferred. Deer may browse the leaves if other food is scarce, but damage is infrequent and the plant recovers quickly. Grows 5–6 feet tall, 4–5 feet wide, but can be trimmed to a more compact shape for container use. Zones 7–9. Deer resistance: B.

MISS LEMON ABELIA (*Abelia* ×*grandiflora* 'Miss Lemon').

Evergreen in milder climates, with brightly variegated leaves on burgundy stems. Pale pink tubular blooms attract hummingbirds in summer. Grows 3–4 feet tall and wide. Zones 6–9. Deer resistance: B.

PERSIAN SHIELD (*Strobilanthes dyerianus*).

Iridescent purple leaves overlaid with silver and deep green veins make this heat-tolerant plant a favorite of savvy gardeners. Thrives in partial shade, where it will grow to 3 feet tall and 2 feet wide. Enjoy as a houseplant and summer annual in temperate climates. Zones 9–11. Deer resistance: B.

How do you successfully combine several different patterns in one container without it looking like Grandma's crazy quilt? The trick is to start with one key plant, in this case the bold, tropical-looking fatshedera. With very similarly colored (albeit smaller) leaves, the abelia easily slips in as a secondary player. Adding notable contrast and running as a river between the two is the heavily veined purple leaf of Persian shield. Site: partial shade.

Contemporary Meets Classic

PLATINUM BEAUTY LOMANDRA
(*Lomandra longifolia* 'Platinum Beauty'). This shimmery, grass-like plant features narrow blades of green and white striped foliage that move in the breeze. Drought tolerant once established. Grows 2–3 feet tall and wide. Zones 8–10. Deer resistance: A.

ANGEL WINGS SENECIO
(*Senecio candicans* 'Angel Wings'). Dramatic, silver, heart-shaped foliage is truly stunning in containers and has proven to be both drought and salt tolerant. To 10–12 inches tall and wide. Zones 7–10. Deer resistance: A.

QUICKSILVER HEBE
(*Hebe pimeleoides* 'Quicksilver'). The wiry black stems of this low-growing, airy shrub are studded with tiny blue-gray leaves. Pale lavender flowers appear in summer. Like all hebes, tolerant of wind and drought. Grows 1 foot tall, 2–3 feet wide. Zones 8–11. Deer resistance: B.

KIRIGAMI OREGANO
(*Origanum* ×*hybrida* 'Kirigami'). An ornamental spiller for containers, with fragrant, blue-green foliage and hop-like rosy-lavender bracts. Grows 8–10 inches tall and 12–14 inches wide. Drought tolerant once established. Zones 5–8. Deer resistance: A.

This sculptural container, which itself serves as a bold focal point, called for a restrained planting that allowed its clean lines to remain visible. The classic black and white color scheme, updated with the addition of silver and rose, focuses on contrasting foliage textures and connects to the surrounding landscape. Site: full sun.

RESOURCES

Introduction
Designers: Barbara Katz (londonlandscapes.com), Arcadia Gardens LLC (arcadiagardensllc.com)
Landscape architect: Tait Moring & Associates (taitmoring.com)

A Designer's Dream Garden
Designer: Le Jardinet (lejardinetdesigns.com)
Contractor: Berg's Landscaping (bergslandscaping.com)
Artist: Jesse Kelly (jessekellyglass.com)
Vegetable garden plans: Le Jardinet (lejardinetdesigns.com/critter-proof-vegetable-garden-design-plans)
Products: Liquid Fence (liquidfence.com), Plantskydd (plantskydd.com)

A Country Garden
Architect: Francis C. Klein & Associates (francisclein.com)
Designer: Susan Cohan Gardens (susancohangardens.com)
Plant specialist: Peony's Envy (peonysenvy.com)

A Desert Oasis
Inspiration: Pam's blog, *Digging* (penick.net/digging)
Artists: TerraTrellis (terratrellis.com), Red Grass Designs (etsy.com/shop/redgrassdesigns)

A Blue Jeans Garden
Designer and contractor: Arcadia Gardens LLC (arcadiagardensllc.com)
Product: Liquid Fence (liquidfence.com)

A Storyteller's Garden
Designer and contractor: Sifford Garden Design (siffordgardendesign.com)
Artists: Kimberly Tyrrell (KTyrrell@clayworksinc.org), Jesse Kelly (jessekellyglass.com), Pavel Efremoff (pafa.org/faculty-members/pavel-efremoff), Benjamin Parrish (steeldesignstudios.com), Jim Weitzel (weitzelart.net)
Product: Deer Out (deerout.com)

A Garden of Survivors
Designer: Maria Smithburg, Artemisia Landscape Design (maria@msartemisia.com)
Artist: Todd Brooks, Arcadia Gardens LLC (arcadiagardensllc.com)
Product: Liquid Fence (liquidfence.com)

A Garden of Connections
Architect: Michael McCulloch AIA (Mike@mmarch.com)
Water feature designer: Eamonn Hughes (hugheswatergardens.com)
Landscape design team: Ann Lovejoy (gardeningwithannlovejoy@gmail.com), Beth Holland (tallwoodsgardenhouse.com), Laura Crockett (Laura@GardenDiva.com), John Greenlee (greenleeandassociates.com)
Artist: Lee Kelly (lee-kelly.net)
Bulk seed: Silver Falls Seed Company (silverfallsseed.com)

A Confetti Garden

Designer: Diana's Designs (dianasdesignsaustin.com)

Inspiration: Diana's blog, *Sharing Nature's Garden* (dianasdesignsaustin.com/category/sharing-natures-garden); *Central Texas Gardener* (centraltexasgardener.org); Lady Bird Johnson Wildflower Center (wildflower.org)

A Lake House Garden

Designer and contractor: Anna Brooks, Arcadia Gardens LLC (arcadiagardensllc.com)

A Collector's Garden

Inspiration: Western New York Hosta Society (wnyhosta.com), Gardens Buffalo Niagara (gardensbuffaloniagara.com), The Hosta Lists (hostalists.org), Hosta Library (hostalibrary.org)

Products: Milorganite (milorganite.com), Bobbex (bobbex.com)

A Hilltop Hacienda

Architect: Charles Travis AIA (chasarchitects.com)

Landscape architect: Tait Moring & Associates (taitmoring.com)

Inspiration: Lady Bird Johnson Wildflower Center (wildflower.org)

Product: Habiturf (wildflower.org/learn/how-to/create-a-native-habiturf-lawn)

A Curated Garden

Artist: Georgia Gerber (georgiagerber.com)

Inspiration: Bloedel Reserve (bloedelreserve.org), Heronswood (heronswoodgarden.org), Stanley Park (vancouver.ca/parks-recreation-culture/stanley-park.aspx)

Product: Bobbex (bobbex.com)

A Suburban Retreat

Designer: Susan Cohan Gardens (susancohangardens.com)

Contractor: Frank Scheppe (scheppelandscape.com)

Products: Liquid Fence (liquidfence.com), Deer Off (havahart.com/about-deer-off)

Deer-Resistant Container Gardens

Designers: Pam Penick, for "Whales on Wheels," "Wild Tresses," and "Sparks May Fly"; Judy Muntz, for "Banana and Blueberry Sundae" and "Tropical Romance"; Tait Moring & Associates (taitmoring.com), for "Snake Charmer." All other designs by Le Jardinet (lejardinetdesigns.com).

METRIC CONVERSIONS

INCHES	CM
1	2.5
2	5.1
3	7.6
4	10
5	13
6	15
7	18
8	20
9	23
10	25
20	51
30	76
40	100
50	130

FEET	M
1	0.3
2	0.6
3	0.9
4	1.2
5	1.5
6	1.8
7	2.1
8	2.4
9	2.7
10	3
20	6
30	9
40	12
50	15

TEMPERATURES
$°C = 5/9 \times (°F - 32)$
$°F = (9/5 \times °C) + 32$

FURTHER READING

Chapman, Karen, and Christina Salwitz. 2013. *Fine Foliage*. St. Lynn's Press, Pittsburgh, PA.

———. 2017. *Gardening with Foliage First*. Timber Press, Portland, OR.

Clausen, Ruth Rogers. 2011. *50 Beautiful Deer-Resistant Plants*. Timber Press, Portland, OR.

Ferguson, Nicola. 2005. *Right Plant, Right Place*. Fireside, New York, NY.

Hart, Rhonda Massingham, and Jim Wilson. 2005. *Deerproofing Your Yard and Garden*. Storey Publishing, North Adams, MA.

Singer, Carolyn. 2006. *Deer in My Garden*. Vol. 1, *Perennials and Subshrubs*. Garden Wisdom Press, Grass Valley, CA.

Soderstrom, Neil. 2009. *Deer-Resistant Landscaping*. Rodale, New York, NY.

ACKNOWLEDGMENTS

I would love to know how anyone manages to run a business, take care of a family and pets, tend their garden, and do basic things like laundry while also writing a book. I'm pretty good at multitasking but that level of juggling is clearly beyond my capabilities, so my greatest thanks must go to my amazing husband Andy, who managed to catch all the balls I dropped and still remembered to bring me wine and chocolate on a regular basis as I worked late into the night. I'm so blessed to share my life with you.

When our two children were small I took great delight in sharing my love of the outdoors with them; picking sun-kissed strawberries, making snapdragons "talk," jumping in puddles on rainy days. While Paul's interest in gardening appears to be limited to what he can find to eat in it, our daughter Katie now tends a lovely garden of her own, passing her knowledge on to her own infant daughter, to whom this book is dedicated. I am both delighted and immensely proud that Katie has produced all the charming illustrations for this book; her creativity and attention to detail have added an artistic flair to these pages that I could never have achieved alone, and it is richer for her contribution.

To the wonderful homeowners who graciously allowed me into their lives, homes, and gardens, often at unsociable hours—thank you. Your tenacity as deer-plagued gardeners is matched only by your generous spirit and hospitality. I am honored to have the opportunity to share your stories.

It took a network of friends, designers, fellow writers, and even complete strangers to discover the gardens showcased here. Then having found them, I often needed help to identify mystery plants or understand nuances of climates and soils I was unfamiliar with. I would therefore especially like to acknowledge the following people, who helped me so tirelessly with this important research and answered my endless questions: Anna Brooks, Susan Cohan, Connie Cottingham, Sally Cunningham, John Greenlee, Beth Holland, Judith Jones, Barbara Katz, Linda Lehmusvirta, Elizabeth McGreevy, Tait Moring, Kelly Norris, Thad Orr, Pam Penick, Jenny Peterson, Mike Shadrack, and Richie Steffen.

Finally, my profound gratitude and thanks to the talented team at Timber Press for translating my ideas into this beautiful book: Tom Fischer, senior acquisitions editor, who has been my advocate and cheerleader from the beginning; photo editor Sarah Milhollin, who worked tirelessly with me for eighteen months, reviewing digital files, offering photography tips, and encouraging me as I endeavored to edit thousands of images; and project editor Franni Farrell, whose remarkable editing skills are matched by her enthusiasm and sense of humor. Together we have created a book that we can all be immensely proud of.

Now if only the deer would read it.

PHOTO CREDITS

Todd Brooks, page 69 (top right).
Janet Davis (thepaintboxgarden.com), page 203 (second from top).
Deborah Heg, page 190 (middle).
Diana Kirby, page 128 (top).
Pam Penick, pages 55 (top left) and 60 (top left).
Christina Salwitz, pages 50 (third from top) and 76 (top).
Jay Sifford, page 88 (bottom).
All other photos are by the author.

INDEX

A
abelia, 133, 226
Abelia ×*grandiflora*, 226
Abyssinian banana, 210
Acapulco wedelia, 129, 135
Acer circinatum, 186
Acer grandidentatum, 168, 174
Acer griseum, 24
Acer macrophyllum, 20, 178, 179
Acer palmatum, 27, 33, 85, 94, 148, 201
Acer pseudoplatanus, 81
Acer rubrum, 28
Acer saccharinum, 154
Acer saccharum, 174
Achillea, 50
Achillea millefolium, 22, 35, 115, 119
Aconitum carmichaelii, 199, 203
Actaea simplex, 103, 105, 199
Actinidia kolomikta, 100
Adam's needle, 102
Adenocarpus decorticans, 177, 190
Adiantum pedatum, 150
Adiantum venustum, 209
Aegopodium podagraria, 32
African daisy, 224
agave, 168, 186
Agave americana, 171, 173, 187
Agave ovatifolia, 129, 210
Agave tequilana, 171
Ageratum houstonianum, 220
Ajuga reptans, 31
Alchemilla mollis, 101, 102, 106
Allium, 34, 101
Allium ampeloprasum, 116
Allium schoenoprasum, 102
allspice, 51
Alocasia macrorrhizos, 217
alpine mint bush, 187
Alternanthera ficoidea, 220
Amelanchier, 183
American beautyberry, 134, 137
American cranberrybush, 69
American hornbeam, 148, 151
Amsonia hubrichtii, 27, 28, 37, 46
Anisacanthus quadrifidus var. *wrightii*, 169
Anthurium, 209
Antirrhinum majus, 128, 129
Aquilegia coerulea, 198
Aquilegia vulgaris, 69, 76
Arbutus menziesii, 183
Arctostaphylos densiflora, 181
Arctostaphylos uva-ursi, 186
Arkansas blue star, 27, 28, 37, 46
aromatic aster, 68
Aronia melanocarpa, 112
Aruncus aethusifolius, 159
Aruncus dioicus, 203
Asclepias curassavica, 129
Asparagus densiflorus, 58, 214
aspen, 28
astelia, 209
astilbe, 69, 159, 199
Astilbe chinensis, 149
Astilbe chinensis var. *pumila*, 161
Astilbe simplicifolia, 104
Athyrium, 88, 209
Athyrium filix-femina, 159, 161
Athyrium filix-femina var. *angustum*, 202, 204
Athyrium niponicum var. *pictum*, 70, 88, 104
Atlas cedar, 85, 87
Aucuba japonica, 198
Austroderia richardii, 179
autumn fern, 29, 89, 93, 94, 202
autumn sage, 129, 132, 137
azalea, 26, 199

B
bald cypress, 82, 85, 94
bamboo muhly, 58, 60, 61
banana, 167
Baptisia, 212
Baptisia australis, 46, 68, 74, 76
barberry, 17, 27, 31, 33, 36, 93, 101, 190, 191, 211, 222, 223, 224
barrenwort, 107, 186, 199
basil, 102, 172
beaked yucca, 129
bearberry cotoneaster, 31
bearded iris, 101
beautyberry, 221
beebalm, 46, 115, 116, 121, 159
beech, 80
beefsteak plant, 220
begonia, 209, 216
Begonia boliviensis, 209, 216
Berberis, 101
Berberis thunbergii, 27, 31, 33, 36, 93, 191, 211, 223, 224
Berberis wilsoniae, 190
Bergenia cordifolia, 188
Berkeley sedge, 56
Berlandier's sundrops, 129
Beschorneria yuccoides, 129
betony, 46, 50
Betula nigra, 28, 48
Betula pendula, 154, 155
Betula utilis var. *jacquemontii*, 24
bidens, 211
big blue lilyturf, 69
bigleaf maple, 20, 178, 179
bigtooth maple, 168, 174
bishop's weed, 32
black bugbane, 105, 199
black-eyed Susan, 27, 69, 159
blackfoot daisy, 175
blanketflower, 119, 121, 219
bleeding heart, 30, 160, 162
Bletilla striata, 133
bluebeard, 33, 221
blue false indigo, 74
blue fescue, 211
blue grama, 168
blue nest spruce, 187
blue nolina, 171
Boston fern, 196

Bouteloua dactyloides, 168
Bouteloua gracilis, 168
boxwood, 42, 49, 91
bridalwreath spirea, 71, 73
bromeliad, 58, 208
Brunnera macrophylla, 148, 149, 159, 217
buckthorn, 224
Buddleja, 220
Buddleja davidii, 143
buffalo grass, 168
bugbane, 103, 105, 199
bugleweed, 30, 31
Bupleurum fruticosum, 179, 189
bur oak, 129
bush cinquefoil, 221
butterfly bush, 143, 220
Buxus, 42, 49

C

Calibrachoa, 210
California lilac, 221, 222
California wax myrtle, 177
Callicarpa, 221
Callicarpa americana, 134, 137
Calycanthus ×*raulstonii*, 51
Calylophus berlandieri, 129
Canadian hemlock, 33, 69, 71, 142
canna lily, 219
Carex, 21, 181
Carex divulsa, 56
Carex oshimensis, 89, 95
Carex phyllocephala, 58, 133
Carex retroflexa, 56
Carex sideroticha, 103
Carpinus caroliniana, 148, 151
Caryopteris ×*clandonensis*, 33, 221
castor oil plant, 17
catmint, 27, 46, 76, 101, 143, 150
Ceanothus griseus var. *horizontalis*, 222
Cedrus atlantica, 85, 87
Cedrus deodara, 24, 95, 187

Cedrus lawsoniana, 24
celosia, 221
Celosia argentea var. *cristata*, 221
century plant, 171, 173, 187
Cephalotaxus harringtonia, 60, 62, 132
Cercidiphyllum japonicum, 28
Cercis canadensis, 68, 196, 199, 203
Chaenomeles speciosa, 157
Chamaecyparis obtusa, 24
Chamaerops humilis var. *argentea*, 58, 62
Chasmanthium latifolium, 60
Cheddar pinks, 34, 87, 131
Chilean mesquite, 169
Chinese astilbe, 149, 161
Chinese fringe flower, 132, 221, 222, 223
Chinese ground orchid, 133
Chinese mahonia, 132
Chinese wisteria, 144, 150
chinquapin oak, 168
chives, 102
chokeberry, 112
Christmas fern, 89
clematis, 100
Cleome, 220
climbing hydrangea, 100
Codiaeum variegatum, 208
Colorado blue columbine, 198
Colorado spruce, 198
columbine, 69, 76, 102
comfrey, 186, 187
common fleabane, 119
coneflower, 101, 103, 154, 159
Conoclinium greggii, 60, 61, 63
Convallaria majalis, 105
Coprosma, 212, 219
coral bells, 84, 102, 146, 147, 158, 201, 202
Coreopsis, 113, 223
Coreopsis verticillata, 24, 77, 143
Cornus, 29

Cornus florida, 202
Cotinus coggygria, 27, 81, 93, 155
Cotoneaster dammeri, 31
cottonwood, 104
cowslip, 31
crapemyrtle, 127
creeping broadleaf sedge, 103
creeping Jenny, 208, 216, 219, 224
creeping thyme, 24, 183
creeping zinnia, 223
crocosmia, 24, 219
croton, 208
Cryptomeria japonica, 187
Cuban buttercup, 129
Cupressus ×*leylandii*, 12
Cupressus sempervirens, 118
curly mesquite, 168
cushion spurge, 106
Cycas revoluta, 129
cypress vine, 132, 135

D

daffodil, 26, 27, 34, 68, 69, 163, 198
dame's rocket, 70–71
Daphne odora, 218
Dasylirion texanum, 62
Dasylirion wheeleri, 129
daylily, 27, 69, 101, 115, 116, 153–160
deadnettle, 217
Dennstaedtia punctilobula, 49
deodar cedar, 24, 95, 187
Deutzia gracilis, 50, 69
Dianella, 215
Dianella tasmanica, 57
Dianthus gratianopolitanus, 34, 87, 131
Dichondra argentea, 60, 211, 224
dogwood, 202
doublefile viburnum, 44, 100, 103, 151
Douglas fir, 118, 178
Drimys winteri var. *chiloense*, 187
drooping fetterbush, 29, 186

Dryopteris erythrosora, 29, 89, 94, 202
Duranta erecta, 134
dwarf Chinese astilbe, 161
dwarf goat's beard, 159
dwarf palmetto, 58, 60
dyckia, 58
Dyckia choristaminea, 214

E
eastern redbud, 68, 196, 199, 203
Echinacea, 101, 154
Echinacea purpurea, 119, 161, 222
Echinocactus grusonii, 168
Elaeagnus ×*ebbingei*, 184, 190
Elaeagnus pungens, 181
elderberry, 219, 223
elephant garlic, 116
English bluebells, 31, 34, 37
English lavender, 122
English primrose, 30
English yew, 187
Ensete ventricosum, 210
Epimedium, 186
Epimedium ×*versicolor*, 107, 199
Erigeron philadelphicus, 119
Eriobotrya japonica, 131
Eryngium, 32, 37
Eryngium giganteum, 115, 123
esperanza, 129, 131, 132, 135, 171
Euphorbia, 217, 224
Euphorbia ×*martinii*, 212
Euphorbia polychroma, 106
Euphorbia rigida, 58, 169
evening primrose, 119, 123
evergreen huckleberry, 191, 218

F
Fagus grandifolia, 80
false holly, 27
false indigo, 46, 68, 76, 212
×*Fatshedera lizei*, 226
Fatsia japonica, 82, 137, 209

fern, 87–89, 209
Festuca glauca, 211
firebush, 169, 171, 173
firecracker plant, 168, 221
flame acanthus, 169
flame grass, 116
flamingo flower, 209
flax lily, 57, 215
floss flower, 220
flowering currant, 183
flowering quince, 157
forsythia, 68
forsythia sage, 134
fountain grass, 58, 61, 123, 210
foxtail fern, 58, 214
fragrant sumac, 116, 122
Fritsch spirea, 69

G
Gaillardia aristata, 119, 121
Gaillardia ×*grandiflora*, 219
Galium odoratum, 104, 105, 201, 202, 205
Galphimia gracilis, 57, 62, 129
Gaultheria procumbens, 183
Gaultheria shallon, 183
Gaura lindheimeri, 33, 35
geranium, 27, 33, 49, 101, 105
Geranium sanguineum, 66, 75
giant hesperaloe, 60, 167
giant Japanese butterbur, 105
giant taro, 217
goat's beard, 159, 203
golden barrel cactus, 168
golden locust, 24, 28
goldenrod, 68
golden sedge, 60, 89, 91
golden thryallis, 57, 62, 129
gopher plant, 58, 169
grape hyacinth, 69
grapevines, 186
gray-woolly twintip, 55, 56, 61, 167

Gregg's blue mistflower, 60, 61, 63
Guzmania, 208

H
hairy Acapulco wedelia, 129, 135
Hakonechloa macra, 21, 103, 148, 149, 158, 198
Hamamelis ×*intermedia*, 191
Hamamelis vernalis, 144
Hamamelis virginiana, 68
Hamelia patens, 169, 171, 173
hay-scented fern, 49
heavenly bamboo, 17, 132
hebe, 216, 217, 227
Hebe pimeleoides, 216, 227
Helenium, 162
heliotrope, 216
Heliotropium arborescens, 216
hellebore, 29, 49, 186, 201, 202, 218
Helleborus, 29, 49, 186, 201, 202
Helleborus ×*ericsmithii*, 218
Helleborus orientalis, 70, 146, 147
Hemerocallis, 27, 69, 101, 154, 158
Hesperaloe funifera, 60, 167
Hesperis matronalis, 70
Heuchera, 84, 158, 201, 202
Heuchera villosa, 102, 146, 147, 201
Hibiscus syriacus, 33, 163, 220
Hilaria belangeri, 168
Himalayan birch, 24, 28, 33
Himalayan maidenhair fern, 209
Hinoki cypress, 24
holly, 27
hosta, 13, 29, 100, 104, 147, 153–160, 184
Hyacinthoides non-scripta, 31, 37
hydrangea, 143, 148
Hydrangea anomala subsp. *petiolaris*, 100
Hydrangea arborescens, 143
Hydrangea macrophylla, 148
Hydrangea quercifolia, 101, 198
Hypericum, 158

I

Ilex, 27
Ilex vomitoria, 133, 168, 175
Illicium parviflorum, 95
Impatiens hawkeri, 147–148
inland sea oats, 60
insideout flower, 187
Ipomoea batatas, 217
Ipomoea quamoclit, 132, 135
iris, 75, 77, 102, 103, 199
Iris sibirica, 101
Italian cypress, 118
Itea virginica, 30, 201, 203

J

Japanese aralia, 82, 137, 209
Japanese aucuba, 198
Japanese cedar, 187
Japanese forest grass, 21, 103, 148, 149, 158, 198, 199
Japanese kerria, 198, 205
Japanese maple, 27, 28, 33, 85, 87, 94, 148, 180, 201
Japanese painted fern, 70, 88, 104
Japanese plum yew, 60, 62, 132
Japanese shrub mint, 188
Japanese stewartia, 144
Japanese tree lilac, 48, 49
Japanese white pine, 84
Jerusalem sage, 112, 121
Joseph's coat, 220
juniper, 33, 36, 185, 189, 223
Juniperus chinensis, 185
Juniperus conferta, 84, 95
Juniperus ×*pfitzeriana*, 189
Juniperus sabina, 185
Juniperus scopulorum, 84
Juniperus squamata, 33, 36, 223

K

katsura, 28
Kerria japonica, 199, 205
kinnikinnick, 186

kiwi vine, 100
knotweed, 187

L

lady fern, 159, 161, 202, 204
lady's mantle, 101, 102, 106
Lagerstroemia, 127
lamb's ears, 69, 75, 113, 129, 224
Lamium maculatum, 217
Lamprocapnos spectabilis, 30, 160, 162
lantana, 131, 132, 219
Lantana camara, 131, 219
Lantana montevidensis, 167
Lavandula allardii, 215
Lavandula angustifolia, 69, 77, 122
Lavandula ×*intermedia*, 120
lavender, 69, 77, 113, 120, 122, 215
lemon thyme, 212, 220
leopard plant, 69, 75, 103, 104, 106
Leucanthemum ×*superbum*, 114, 115
Leucanthemum vulgare, 102, 114, 121
Leucophyllum frutescens, 171, 173
Leucosceptrum japonicum, 188
Leucothoe fontanesiana, 29, 186
Leyland cypress, 12
Ligularia, 69
Ligularia dentata, 75, 103, 104, 106
Ligustrum sinense, 223
lilac, 44, 68, 71, 144, 151, 157, 162
Lilium, 101
lily-of-the-valley, 105
Lindheimer's muhly, 174
Liquidambar styraciflua, 80
Liriodendron tulipifera, 71, 194
Liriope muscari, 69
live oak, 54, 56, 133, 166, 168
Lobelia erinus, 209, 224
Lobularia maritima, 32, 33
Lomandra longifolia, 220, 227
Lonicera pileata, 205
loquat, 131
Loropetalum chinense, 132, 222, 223

Lotus, 223
Lotus berthelotti, 216, 219
lotus vine, 216, 219, 223
lungwort, 154
lupin, 119
Lupinus latifolius, 119
Lupinus polyphyllus, 119
Lupinus ×*regalis*, 119
Lupinus texensis, 168
Lychnis coronaria, 99
Lysimachia, 216
Lysimachia congestiflora, 219, 224
Lysimachia nummularia, 208
Lysimachia punctata, 102

M

Macleaya cordata, 107
madrone, 183
Magnolia grandiflora, 48
Magnolia ×*soulangeana*, 154
Mahonia, 29, 186
Mahonia bealei, 133
Mahonia fortunei, 132
Malvaviscus arboreus var. *drummondii*, 58, 63, 132
×*Mangave*, 60, 61
manzanita, 181
matilija poppy, 186
Matteuccia struthiopteris, 201, 204
mayapple, 88, 104, 186
Melampodium leucanthum, 175
Mexican bush sage, 57, 132, 136
Mexican feather grass, 58, 135, 212
Mexican lily, 129
Mexican petunia, 210
Microbiota decussata, 116, 122, 186
million bells, 210
mint, 32
mirror plant, 212, 219
miscanthus, 21, 46, 116
Miscanthus sinensis, 21, 44, 73
mock orange, 157, 183
Monarda, 46, 115, 121, 159

mondo grass, 91, 186, 218
monkshood, 199, 203
Morella californica, 177
mother-in-law's tongue, 168
mountain holly, 189
Muhlenbergia, 168
Muhlenbergia capillaris, 167
Muhlenbergia dumosa, 58, 61
Muhlenbergia lindheimeri, 174
muhly grass, 58, 167, 168, 171
Musa, 167
Muscari armeniacum, 69

N
Nandina domestica, 17, 132
Narcissus, 27, 163
narrow-leaved zinnia, 129
Nassella tenuissima, 58, 135, 212
nasturtium, 115
Nepeta, 27, 101
Nepeta ×*faassenii*, 143, 150
Nepeta racemosa, 46, 76
Nephrolepis exaltata, 196
New Guinea impatiens, 147
New Zealand holly, 185, 189
Nolina nelsoni, 171
northern lady fern, 202, 204
northern maidenhair fern, 150
Norway spruce, 84, 87, 185
Nuttall's oak, 129

O
oakleaf hydrangea, 101, 198
Ocimum basilicum, 172
Oenothera fruticosa, 119, 123
Olearia ilicifolia, 189
Olearia macrodonta, 185, 189
Onoclea sensibilis, 104, 105
Ophiopogon japonicus, 91
Ophiopogon planiscapus, 186, 218
Opuntia ellisiana, 58, 169
oregano, 102, 211, 215, 227
Oregon grape, 29, 186

oriental hellebore, 70, 146, 147
oriental spruce, 24, 84, 94
Origanum, 215
Origanum ×*hybrida*, 227
Origanum vulgare, 211
ornamental onion, 34, 101, 116
Osmanthus heterophyllus, 27
Osmunda regalis, 188
Osteospermum, 224
ostrich fern, 201, 204
oxeye daisy, 102, 114, 121
Ozark witch hazel, 144

P
Pachysandra terminalis, 51, 103, 201, 202
Paeonia, 81, 149, 163
Paeonia lactiflora, 43
pale-leaf yucca, 55, 63
pale pavonia, 58, 63
Panicum virgatum, 21
Papaver rhoeas, 119
paperbark maple, 24
Parrotia persica, 24, 36, 180
parsnip, 34–35
Parthenocissus quinquefolia, 116
Pavonia hastata, 58, 63
Pennisetum orientale, 123
Pennisetum purpureum, 58, 61
Pennisetum setaceum, 210
penstemon, 51
peony, 34, 43, 81, 101, 149, 163
Perilla, 220
Perovskia atriplicifolia, 68, 69, 116
Persian ironwood, 24, 36, 180
Persian shield, 132, 226
Persicaria virginiana, 187
Petasites japonicus var. *giganteus*, 105
petunia, 210
Phalaris arundinacea, 28
Philadelphus, 157, 183
Phlomis russeliana, 112–113, 121

Picea, 46, 154
Picea abies, 84, 87, 185
Picea mariana, 187
Picea orientalis, 24, 84, 94
Picea pungens, 198
pigsqueak, 188
pine, 33, 154
pink skullcap, 131, 136
Pinus, 154
Pinus parviflora, 84
Pinus strobus, 24
Piper auritum, 132
Platanus occidentalis, 48
plateau goldeneye, 173
plume poppy, 107
Podocarpus macrophyllus, 91, 129
Podophyllum, 186
Podophyllum peltatum, 88, 104
Polygonatum odoratum var. *pluriflorum*, 104, 201, 204
Polystichum acrostichoides, 89
Polystichum polyblepharum, 89
Polystichum setiferum, 188, 218
pomegranate, 131
Populus, 104
Populus tremula, 28
Port Orford cedar, 24
Portuguese laurel, 185
post oak, 129
potato vine, 224
Potentilla fruticosa, 221
primrose, 31
Primula veris, 31
Primula vulgaris, 30–31
privet, 223
privet honeysuckle, 205
Prosopis chilensis, 169
Prostanthera cuneata, 187
Prunus ×*cistena*, 69
Prunus lusitanica, 185
Pseudotsuga menziesii, 118, 178
Pulmonaria, 154
Punica granatum, 131

purple coneflower, 119, 161, 222
purple-leaf sand cherry, 69
purple queen, 212, 216, 220
Pyrus salicifolia, 24

Q
Quercus buckleyi, 168
Quercus coccinea, 186
Quercus macrocarpa, 129
Quercus muehlenbergii, 168
Quercus stellata, 129
Quercus texana, 129
Quercus virginiana, 54, 133, 166

R
red corn poppy, 119
red maple, 28
reed canarygrass, 28
Rhamnus frangula, 224
Rheum rhabarbarum, 101
rhododendron, 68, 104, 186, 199
rhubarb, 69, 101
Rhus aromatica, 116, 122
Rhus typhina, 216
Ribes sanguineum, 183
Ricinus communis, 17
river birch, 28, 46
Robinia pseudoacacia, 24
Rocky Mountain juniper, 84
Rodgers' flower, 29, 105
Rodgersia podophylla, 29, 105
Romneya coulteri, 186
root beer plant, 132
Rosa rugosa, 99, 100
rose, 68, 116, 143
rose campion, 99
rosemary, 58, 172, 174, 220
rose of Sharon, 33, 163, 220
Rosmarinus officinalis, 58, 174, 220
royal fern, 188
Rudbeckia, 69
Rudbeckia fulgida var. *sullivantii*, 27, 159

Ruellia simplex, 210
rugosa rose, 99, 100
Russelia equisetiformis, 168, 221
Russian sage, 68, 69, 116

S
Sabal minor, 58
sage, 44, 46, 51
sago palm, 129
salal, 183
Salvia greggii, 129, 137
Salvia leucantha, 57, 136
Salvia madrensis, 134
Salvia nemorosa, 44
Salvia ×*sylvestris*, 51
Sambucus nigra, 223
Sambucus racemosa, 219
Sansevieria trifasciata, 168, 222
Sanvitalia procumbens, 223
Sarcococca, 186
Sarcococca ruscifolia, 187
saucer magnolia, 154
Saxifraga stolonifera, 82
scarlet oak, 186
Scutellaria suffrutescens, 131, 136
sea holly, 32, 37, 115, 116, 123
sedge, 21, 56, 58, 95, 133, 181
Sedum, 99
Senecio candicans, 212, 227
Senecio jacobaea, 113, 114
sensitive fern, 104, 105
serviceberry, 183
Shasta daisy, 114, 115
shore juniper, 84, 87, 95
shrubby hare's-ear, 179, 189
Siberian bugloss, 148, 149, 159, 217
Siberian cypress, 116, 122, 186
Siberian iris, 77, 101, 102, 199
silverberry, 181, 184, 190
silver birch, 154, 155
silver broom, 177, 190
silver maple, 154
silver Mediterranean fan palm, 58, 62

skyflower, 134
slender deutzia, 50, 69
smokebush, 27, 81, 93, 155
snapdragon, 128, 129
sneezeweed, 162
snowball viburnum, 183
snowberry, 30, 31, 36
society garlic, 131, 132, 136
soft shield fern, 188, 218
Solanum jasminoides, 224
Solidago rugosa, 68
Solomon's seal, 104, 201, 204
Sophora secundiflora, 128, 168, 175
southern magnolia, 48
spider flower, 220
spineless prickly pear, 58, 169
Spiraea fritschiana, 69
Spiraea japonica, 37, 40, 50
Spiraea ×*vanhouttei*, 71, 73
spirea, 37, 40, 50, 68
spruce, 46, 154
spurge, 212, 217, 224
Stachys, 224
Stachys byzantina, 69, 75, 113, 129
Stachys officinalis, 46, 50
star astilbe, 104
Stemodia lanata, 55, 56, 167
Stewartia japonica, 144
St. John's wort, 158
stonecrop, 99
strawberry, 69
strawberry saxifrage, 82
Strobilanthes dyerianus, 132, 226
sugar maple, 174
sumac, 216
sweet alyssum, 32, 33
sweetbox, 186, 187
sweetgum, 80
sweet potato vine, 217
sweet woodruff, 104, 105, 201, 202, 205
switchgrass, 21
sycamore, 48

sycamore maple, 81
Symphoricarpos ×*chenaultii*, 30, 31, 36
Symphyotrichum oblongifolium, 68
Symphytum, 186
Symphytum ×*uplandicum*, 187
Syringa pubescens subsp. *patula*, 144, 151
Syringa reticulata, 48, 49
Syringa vulgaris, 162

T
tansy ragwort, 113, 114
tassel fern, 89
Taxodium distichum, 82, 85, 94
Taxus baccata, 187
Tecoma, 129, 171
Tecoma stans, 131, 132, 135
tequila agave, 171
Teucrium chamaedrys, 188
Texas bluebonnet, 168
Texas mountain laurel, 128, 168, 175
Texas red oak, 168
Texas sage, 171, 173
Texas sotol, 62
Thuja plicata, 69, 178
Thymus citriodorus, 212, 220
Thymus serpyllum, 24, 183
tickseed, 24, 68, 77, 113, 143, 223
toe-toe grass, 179

Tradescantia pallida, 212, 216, 220
trailing lantana, 167
Tropaeolum, 115
tropical milkweed, 129
Tsuga canadensis, 33, 69, 142
Tulbaghia violacea, 131, 136
tulip tree, 71, 194, 201
Turk's cap, 58, 63, 132
Turnera ulmifolia, 129
twig dogwood, 29

V
Vaccinium ovatum, 191, 218
Vancouveria hexandra, 187
verbena, 212, 222
Verbena bonariensis, 222
Viburnum opulus, 183
Viburnum plicatum f. *tomentosum*, 46, 100, 103, 151
Viburnum trilobum, 69–70
Viguiera dentata, 173
vine maple, 186
Virginia creeper, 116
Virginia sweetspire, 30, 201, 203
Vitis, 186

W
wall germander, 188
Wedelia acapulcensis var. *hispida*, 129, 135

Weigela florida, 33, 102, 106, 107
western red cedar, 69, 178
whale's tongue agave, 129, 210
Wheeler sotol, 129
white pine, 24
willowleaf pear, 24
winter daphne, 218
wintergreen, 183
Winter's bark, 187
wisteria, 32, 100
Wisteria frutescens, 144, 150
witch hazel, 68, 69, 191

Y
yarrow, 22, 35, 50, 115, 119
yaupon holly, 133, 168, 175
yellow anise, 95
yellow barrenwort, 107
yellow loosestrife, 102
yew plum pine, 91, 129
yucca, 58, 116, 129, 168
Yucca filamentosa, 102, 129
Yucca pallida, 55, 63
Yucca rostrata, 129

Z
Zinnia angustifolia, 129
Zinnia elegans, 131

Karen Chapman established her award-winning landscape design business, Le Jardinet, in 2006. She is co-author of two books, *Fine Foliage* (St. Lynn's Press, 2013) and *Gardening with Foliage First* (Timber Press, 2017), and her articles and designs have been featured in many online and print publications, including *Garden Design*, *Fine Gardening*, and *Better Homes & Gardens*. Karen also shares her knowledge and passion for gardening through a number of online courses and is a popular speaker at flower and garden shows, botanical gardens, nurseries, and garden clubs across the United States. She lives and gardens on five rural acres in Duvall, Washington, where she is still trying to outwit the deer. Visit her at lejardinetdesigns.com.